THE BILE ACIDS

Chemistry, Physiology, and Metabolism

VOLUME 3: PATHOPHYSIOLOGY

THE BILE ACIDS

A Continuation Order Plan is available for this series. A continuation order will bring delivery of each new volume immediately upon publication. Volumes are billed only upon actual shipment. For further information please contact the publisher.

THE BILE ACIDS

Chemistry, Physiology, and Metabolism

VOLUME 3: PATHOPHYSIOLOGY

Edited by

Padmanabhan P. Nair
Sinai Hospital of Baltimore, Inc.
Baltimore, Maryland

and

David Kritchevsky
The Wistar Institute
Philadelphia, Pennsylvania

PLENUM PRESS • NEW YORK AND LONDON

Library of Congress Cataloging in Publication Data

Many entry under title:

The Bile acids.

Includes bibliographical references.
CONTENTS: v. 1. Chemistry.—v. 2. Physiology and Metabolism.—v. 3. Pathophysiology.
1. Bile acids. 2. Bile salts. I. Nair, Padmanabhan, P., 1931-   ed. II. Kritchevsky, David, 1920-   ed.
QP752.B54B54   599'.01'9243   71-138520
ISBN 0-306-37133-2

© 1976 Plenum Press, New York
A Division of Plenum Publishing Corporation
227 West 17th Street, New York, N.Y. 10011

Printed in the United States of America

CONTRIBUTORS

George Bonorris
Section of Gastroenterology
Department of Medicine
Cedars-Sinai Medical Center
Los Angeles, California

Robert L. Campbell
Department of Surgery
Wayne State University School of Medicine
Detroit, Michigan

Jacqueline Dupont
Department of Food Science and Nutrition
Colorado State University
Fort Collins, Colorado

Helmut Greim
Abteilung für Toxikologie
Gesellschaft für
*   Strahlen-und-Umweltforschung*
Neuherberg, Germany

M. J. Hill
Bacterial Metabolism Research Laboratory
Rear of Colindale Hospital
London, England

Phyllis Janson
Department of Food Science and Nutrition
Colorado State University
Fort Collins, Colorado

David Kritchevsky
The Wistar Institute of Anatomy
*   and Biology*
Philadelphia, Pennsylvania

Jay W. Marks
Section of Gastroenterology
Department of Medicine
Cedars-Sinai Medical Center
Los Angeles, California

E. H. Mosbach
Department of Lipid Research
The Public Health Research Institute
of the City of New York, Inc.
New York City, New York

P. P. Nair
Biochemistry Research Division
Department of Medicine
Sinai Hospital of Baltimore, Inc.
Baltimore, Maryland

Harold J. Nicholas
Institute of Medical Education and Research
*   and Department of Biochemistry*
St. Louis University School of Medicine
St. Louis, Missouri

Norman D. Nigro
Department of Surgery
Wayne State University School of Medicine
Detroit, Michigan

Suk Yon Oh
Department of Food Science and Nutrition
Colorado State University
Fort Collins, Colorado

v

G. Salen
College of Medicine and Dentistry
  of New Jersey
New Jersey Medical School
Newark, New Jersey and
East Orange Veterans Administration
  Hospital
East Orange, New Jersey

Leslie J. Schoenfield
Section of Gastroenterology
Department of Medicine
Cedars-Sinai Medical Center
Los Angeles, California

Jon A. Story
The Wistar Institute of Anatomy and Biology
Philadelphia, Pennsylvania

PREFACE

The first two volumes of this series addressed themselves to the chemistry, physiology, and metabolism of the bile acids. The present volume is devoted to the pathophysiology of bile acids.

As the role of bile acids in health and disease is being increasingly recognized, we have chosen for discussion a wide range of topics of current importance. The presence of bile acids in brain tissue and their possible role in demyelinating diseases form the subject of a provocative discussion. As an extension of this theme, the presence and quantification of bile acids in extrahepatic tissues is the subject of one chapter. The pathophysiological implications of bile acids at the macromolecular level is highlighted by a chapter on the influence of bile salts on the activity of various enzymes.

The general area of hepatobiliary diseases is discussed in two chapters: one describes changes in bile salt metabolism in liver diseases and the other focuses on cholesterol gallstones and their formation and dissolution.

Cerebrotendinous xanthomatosis has been shown to entail a defect in bile acid and sterol metabolism, and this metabolic error is the subject of an illuminating exposition.

There is presently a concerted research effort being brought to bear on the causes of colon cancer, and one important aspect of this work centers on bile acid metabolism. Aspects of bile acid metabolism and cancer are the subject of two chapters. And finally, the role of dietary fiber in bile acid metabolism is updated.

We wish to acknowledge the invaluable assistance of Miss Jane T. Kolimaga in the preparation of the indexes. This work was supported in part by Grants HL-03299, HL-05209, AM-02131 and a Research Career Award (DK), HL-0734 from the National Institutes of Health, United States Public Health Service.

P. P. N.
Baltimore, Maryland

D. K.
Philadelphia, Pennsylvania

CONTENTS

Chapter 4
Bile Acids in Hepato-Biliary Diseases 53
by Helmut Greim

Chapter 7
Bile Acids and Intestinal Cancer **155**
by Norman D. Nigro and Robert L. Campbell

Chapter 8
Fecal Steroids in the Etiology of Large Bowel Cancer **169**
by M. J. Hill

Chapter 9
Dietary Fiber and Bile Acid Metabolism **201**
by David Kritchevsky and Jon A. Story

Chapter 1

BILE ACIDS AND BRAIN

Harold J. Nicholas

*Institute of Medical Education and Research
and Department of Biochemistry
St. Louis University School of Medicine
St. Louis, Missouri*

I. INTRODUCTION*

Some forty-five years ago Weil noted that taurocholic acid caused marked demyelination *in vitro* (1) and proposed that this or some related compound might act as a natural demyelinating agent by entering the brain from the blood stream. The idea that bile acids of some type might be formed biosynthetically within the brain was developed some time later (2) and is the basis for this chapter. There is now considerable evidence that this can occur. This idea in no way displaces evidence that the primary cause of some of the demyelinating diseases may be a virus infection, under genetic control, or due to other causes.

The demyelinating diseases are a group of devastating illnesses of humans and other mammals which may, in the case of multiple sclerosis, terminate in complete debilitation of the individual so afflicted, usually over a period of many years, sometimes within months. The basic defect resulting in paralysis and total disability of the afflicted mammal is essentially due, in the "primary" demyelinating diseases (e.g., multiple sclerosis [MS], Guillain-Barre syndrome, experimental allergic encephalomyelitis [EAE], and diphtheric neuritis) to dissolution of myelin surrounding the myelinated

---

* The following systematic names are given to steroids and bile acids referred to by trivial names: cholestanol, 5α-cholestan-3β-ol; desmosterol, cholesta-5,24-dien-3β-ol; cholic acid, $3\alpha,7\alpha,12\alpha$-trihydroxy-5β-cholanoic acid; chenodeoxycholic acid, $3\alpha,7\alpha$-dihydroxy-5β-cholanoic acid; deoxycholic acid, $3\alpha,12\alpha$-dihydroxy-5β-cholanoic acid; lithocholic acid, 3α-hydroxy-5β-cholanoic acid; isolithocholic acid, 3β-hydroxy-5β-cholanoic acid.

axons of the central or peripheral nervous system. In the "secondary" demyelinating diseases (e.g., Wallerian degeneration and amyotrophic lateral sclerosis [ALS]) myelin disintegrates after destruction of the axon. Since these are, with the exception of EAE, clinical diseases, any further discussion about their initial origin and the immunological aspects would be inappropriate here, and the reader is referred to authoritative references such as (3,4).

The following chapter is based on the hypothesis that, assuming a viral or other primary cause for such disease, the initiating agent may set in motion secondary effects which produce a specific chemical(s) that induces demyelination. Admittedly this is an oversimplification of an extremely complex process, but where such dreadful diseases are involved every aspect of cause should be explored.

Two such chemicals immediately come to mind, since they could potentially be produced or released in large quantities in the central nervous system (CNS) by an aberrant biochemical defect. The first of these is lysolecethin, a powerful hemolytic and demyelinating agent. It is actually present in minute amounts in normal brain tissue (5), and could conceivably be produced in large amounts from any one of a number of phospholipids present in brain. Second, certain metabolic or degradation products of cholesterol are also marked demyelinating agents and have been shown to produce demyelination in central nervous tissue *in vitro* and *in vivo*. Cholesterol itself constitutes about 8% of the dry weight of brain, and about 20–25% of the myelin sheath itself. Recent evidence has indicated that at least certain cholesterol pools of the CNS undergo constant metabolic activity throughout the life of a mammal. Before approaching the question of cholesterol degradation in brain it was felt necessary to present a brief summary of available data substantiating the metabolic activity of CNS cholesterol, both in terms of biosynthesis and turnover.

II. BRAIN CHOLESTEROL BIOSYNTHESIS

Studies on the biosynthesis of brain cholesterol represent one of the most complex areas of sterol biosynthesis, and without doubt, one of the most frequently misinterpreted. For example, Swann *et al.* (6) have made what could be a major observation on brain cholesterol biosynthesis by demonstrating an active cholesterol feedback system in guinea pig tissue such as lung, ileum, and brain. In their publication, however, they make the statement, "Whereas it is well known that the brain of the newborn rodent will synthesize sterol (7), in the rat this process decreases rapidly after birth, and in the adult rat the brain is not capable of significant rates of steroidogenesis" (7,8). In this one sentence the efforts of numerous neurochemists who have pursued this problem over the past 20 years have been completely ignored (see 9–13 for reviews). Swann *et al.* (6) have repeated a concept incorrectly derived from the early work of Waelsch *et al.* (14) and propagated by the observations of Srere *et al.* (7) and their own work. No one experienced in the field of brain cholesterol biosynthesis would expect significant [^{14}C]cholesterol formation by incubating adult rat brain slices for two hours with ^{14}C-labeled acetate. The reasons for this are many and complex, and not completely understood (10). The experience of some 20 years of research involving intraperitoneal (15–17), intracisternal (18), intraventricular (19), and intracerebral (20,21) injection of several ^{14}C-labeled cholesterol precursors has proved without doubt that the adult rat brain retains an active capacity to synthesize cholesterol. It will be noted that the selection of both animal and labeled precursor even for *in vivo* studies is important (15,17,22–25).

In summary, it has been experimentally proven over a period of many years that the capacity of the brain to synthesize cholesterol, whether it be in the rat, mouse, guinea pig, or nonhuman primate (baboon), remains active *throughout the life of the animal.* It may be of more than passing interest to suggest that if this premise applies to a primate like the baboon (25), then it surely must apply to the human as well. It is time that this concept be accepted by the scientific community lest progress in this area be significantly retarded.

III. BRAIN CHOLESTEROL TURNOVER

If cholesterol biosynthesis in the central nervous system* continues throughout the life of an animal, as previously indicated, there must be a continuous turnover of the sterol in this tissue, otherwise a pathological accumulation would occur. Kabara (9) has presented an excellent critique of the intricacies of brain cholesterol "turnover." Despite early conclusions to the contrary (26), there now seems to be general agreement among neurochemists studying this phenomenon that a small but measurable degree of brain cholesterol turnover does occur over long intervals. Davison *et al.* (27) in a relatively long-term study (120 days) concluded that there are at least two metabolic compartments in brain, one in the gray matter and cells

---

* Most of the references presented in this chapter are based on studies of brain tissue, not the spinal cord. In general the metabolic processes for both entities are the same. The differences, where they have been found, are beyond the scope of discussion in this chapter and the reader is referred to the reviews previously suggested for Brain Cholesterol Biosynthesis.

where lipid turnover is rapid, and one in myelin where turnover is very slow or does not occur at all. In any case most experiments since then have shown high concentrations of ^{14}C-labeled cholesterol in brain for long periods following administration. Only a few examples can be given here. Davison *et al.* (28) injected [4-^{14}C]cholesterol into the yolk sac of newly hatched chickens and 220 days later found radioactive cholesterol in the chicken brains, whereas none was detected in liver and plasma. McMillan *et al.* (18) found labeled [^{14}C]cholesterol in rat brains 72 days after intracisternal injection of [2-^{14}C]sodium acetate, and Nicholas and Thomas (21), after one year following intracerebral injection of this same labeled precursor. During the intervening years the problem became more refined and accordingly more complicated. Specific localization in certain areas or compartments of the brain with long-term [^{14}C]cholesterol retention in these areas was indicated by Kritchevsky and Defendi (29), and Khan and Folch-Pi (30). Spohn and Davison (31) suggested that the brain contains a pool of cholesterol with which all membrane structures, including myelin, may readily exchange. In probably the most comprehensive of recent studies [(32,33) and references therein], Chevallier and his colleagues have established that definite and continuous exchange of ^{14}C-labeled brain cholesterol with plasma occurs (32) (and vice versa) and by detailed microscopic radioautography (33) demonstrated cholesterol exchange within all cerebral structures. One of their conclusions is, "Thus, cerebral cholesterol is not an inert compound that is in continual transfer within the brain, likely exchanging between the different cerebral structures" (33). We do not agree with their statement, however, that "The most striking fact revealed by this study is that, once synthesized in the brain, cholesterol disappears entirely in about 20 months" (32). This observation was obtained by extrapolation of decreasing labeled cholesterol specific activities from shorter term experiments. In numerous experiments conducted in the author's laboratory, following intracerebral injection of [2-^{14}C]mevalonic acid into myelinating rats, brain cholesterol remained highly labeled long after (e.g., two years) [^{14}C]cholesterol was depleted from the other organs (34). The difference in results from the two laboratories may be explained on the basis of different amounts of precursor used. The higher the specific activity of the brain cholesterol originally obtained the more ^{14}C-labeled cholesterol will remain for slow turnover. We have recently found that rats so injected continuously excrete into their urine a ^{14}C-labeled metabolite, with none being excreted into the feces (34). While at first it seemed possible that, due to the "[2-^{14}C]mevalonic acid shunt" (35) the compound might be a metabolite coming from some brain lipid other than cholesterol, rats similarly injected with [4-^{14}C]cholesterol also continuously secrete the same material into the urine, but not into feces. It seems possible that the cholesterol-derived compound coming from the brain may be formed in the brain and pass unchanged into the urine. The compound appears to be an unusually stable conjugate of cholesterol or some cholesterol metabolite and is dialyzable. Its structure is currently under active investigation in our laboratory. In other experiments rats injected intraperitoneally with [2-^{14}C]mevalonic acid excrete ^{14}C into the urine, but also pass large amounts of ^{14}C in the feces. The urinary product in this case possibly arises from ^{14}C-labeled cholesterol in other organs and is soon depleted, since brain cholesterol is not heavily labeled following intraperitoneal injection of even large amounts of [2-^{14}C]mevalonic acid (24).

In summary, all evidence points to continuous turnover of brain cholesterol throughout the life of mammals. This, of course, would explain the maintenance of active biosynthetic capacity throughout the same period in order to replace cholesterol lost in the turnover process.

IV. REACTIONS OF NORMAL BRAIN TISSUE WITH THE CHOLESTEROL NUCLEUS

Although cholesterol (I) is the major sterol of brain, a number of minor neutral sterols have been detected in fresh brain tissue (Fig. 1). These include cholestanol (II), 7α-(III) and 7β-hydroxycholesterol (IV), 7-ketocholesterol (V), cholestane-3β,5α,6β-triol (VI), 26-hydroxycholesterol (VII), 25-hydroxycholesterol (VIII), and 24-hydroxycholesterol (IX) (see Van Lier and Smith (36) for review of early work). Desmosterol (X), the immediate precursor of cholesterol, is present in measurable quantity in brain just prior to myelination (37,38) but is virtually absent in the normal adult brain. In addition, cholesterol esters have been found in increased amounts in brain in a variety of pathological demyelinating conditions (39–43).

7α- and 7β-hydroxycholesterols, 7-ketocholesterol, and 24-hydroxycholesterol are generally considered to be auto-oxidation products obtained during the isolation of cholesterol from brain tissue. On the other hand in numerous *in vitro* experiments involving incubations of [4-^{14}C]cholesterol with brain tissue preparations no unusual accumulation of [^{14}C]7α-7β-hydroxycholesterol or [^{14}C]7-ketocholesterol was found, even after prolonged incubation periods (44), and perhaps the question of their origin, biosynthetic or artifact, should be left an open issue. There has as yet, however, been no biochemical evidence presented demonstrating that 7α-hydroxycholesterol can be formed enzymatically from brain tissue. Since the formation of this compound appears to be the initial rate-control-

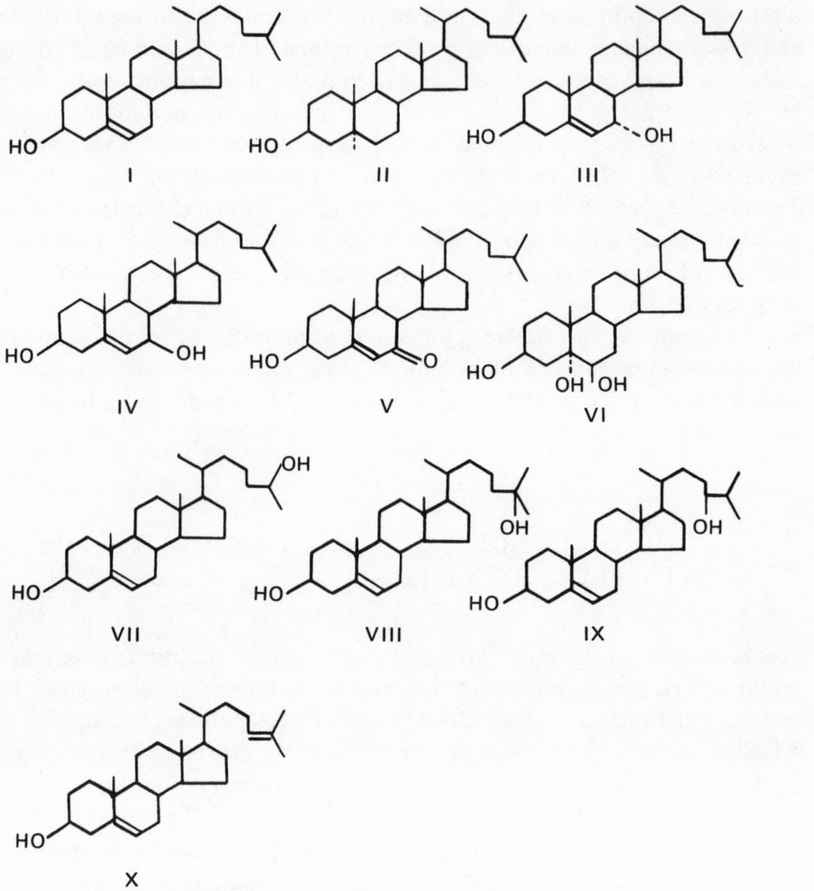

Fig. 1. Minor neutral sterols found in various mammalian brain tissues.

ling step in bile acid biosynthesis in liver (see 45–47 for reviews) this suggests that formation of the classical primary mammalian bile acids such as cholic acid (XI) and chenodeoxycholic acid (XII) (Fig. 2) cannot be formed by brain tissue. It would seem far-fetched that brain tissue, however, would have the capacity to perform such a highly specific step such as 12α-hydroxylation of the sterol nucleus. On the other hand, some reactions which would not normally be anticipated, have been found to occur both *in vitro* and *in vivo*. Thus rodent brain *in vitro* can oxidize 11-hydroxy steroids (48,49), can acetylate cortisol (50), and can metabolize testosterone (51) to more polar compounds of unknown identity. Dehydroepiandrosterone *in vivo* can be converted to ring D hydroxylated δ^5-compounds by brain tissue (52). A recent comprehensive review has extensively surveyed conversion of androgens to estrogens by central nervous tissue (53). The significance of these observations cannot be assessed here, except to note that they also indicate active metabolic conversion of steroids by brain tissue.

Brain tissue preparations can also react with cholesterol to give enzymatic oxidation products involving rings A and B. [4-^{14}C]Cholesteryl palmitate, for example, when incubated with a 12,000g supernate fraction from rat brain, an NADH-generating system, and an antioxidant (β-mercaptoethylamine) is converted to cholestane-3β,5α,6β-triol (VI) and a mixture of [4-^{14}C]cholesterol 5α- and 6α-epoxide (XIII), and [4-^{14}C]cholesterol 5β,6β-epoxide (not shown) (54). At the moment these observations seem to have no bearing on our problem, except to demonstrate that ring B of cholesterol is not inert to attack by brain tissue. This is also evident as indicated by the conversion of [24-^{14}C]lithocholic acid (XIV) to 3α,7β-dihydroxy-5β-cholanoic acid (XV) by adult rat brain cell-free homogenates (55). In the latter work it was also found that the lithocholic acid was oxidized to 3-keto-5β-cholanoic acid (XVI). The reverse process, the conversion of 3-keto-5β-cholanoic acid to lithocholic acid by guinea pig brain preparations (minced or cell-free) has also been shown (56). This reac-

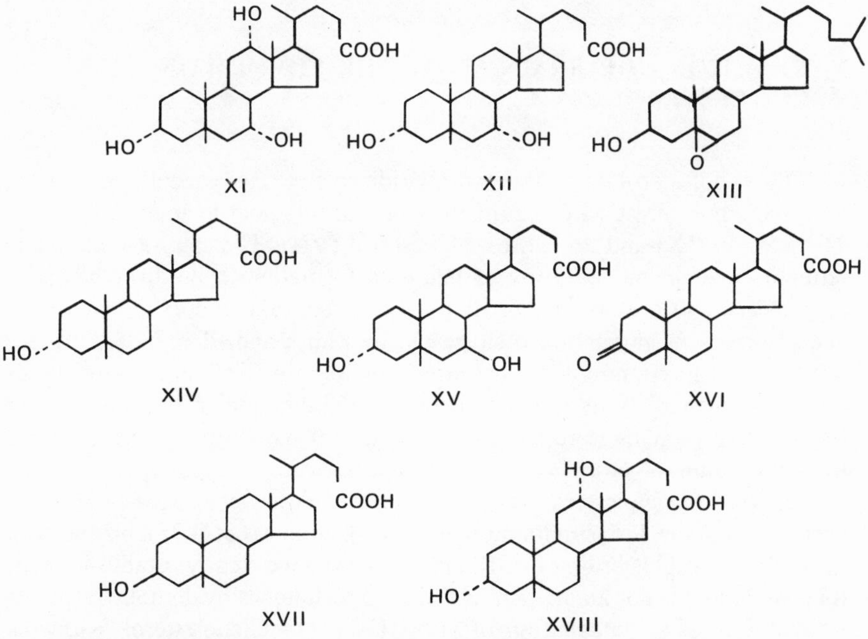

Fig. 2. Acidic steroids which have been detected in or found to be metabolized by mammalian brain tissue *in vitro* or *in vivo*. Cholic acid, XI, is an exception, not having been detected or incubated.

tion required NADH and was stereochemically nonspecific in that both lithocholic acid and isolithocholic acid (XVII) were formed. The conversion of sodium [24-^{14}C]lithocholate by brain to ^{14}C-labeled 3-keto-5β-cholanoic acid following intracerebral injection of the lithocholate has also been shown (57). In the conversion of cholesterol to bile acids in the liver, inversion at C-3 takes place via a 3-keto intermediate (58,59), and reduction of this ketone to the hydroxyl requires NADPH (60). Thus one of the key steps in the conversion of cholesterol to bile acids has been shown capable of being performed by brain tissue (55,56). Martin and Nicholas (61) studied the subcellular localization in adult rat brain of the 3α-hydroxysteroid dehydrogenase responsible for this reaction. The activity was found in the cytosol fraction. The *in vitro* conversion of [24-^{14}C]3-keto-5β-cholanoic acid to lithocholic acid was found to occur without added cofactors. In this same work the conversion of ^{14}C-labeled chenodeoxycholic acid (XII) and deoxycholic acid (XVIII) to their corresponding ^{14}C-labeled 3-keto derivatives was also found to occur on incubation with adult rat brain cell-free preparations, additional evidence for the ability of brain tissue to attack the 3-hydroxyl group of steroids. Very probably guinea pig brain tissue can perform this same step (57).

V. REACTIONS OF BRAIN WITH THE SIDE CHAIN OF CHOLESTEROL

Two cholesterol derivatives with oxidized side chains could be of more than passing interest with regard to potential bile acid formation in brain: 24-hydroxy- (IX) and 26-hydroxycholesterol (VII). The significance of the latter in a secondary pathway of bile acid formation in the liver has been emphasized and reviewed by Javitt and Emerman (62), and its uniqueness of accumulation in human aorta suggested by Smith and Pandya (63). It is a minor sterol in commercially prepared bovine brain and spinal cord lipids (64). Although early attempts to detect the compound in brain and other tissues from humans were not successful (36) it has recently been detected in human brain (65,66).

24-Hydroxycholesterol (cerebrosterol, cholest-5-ene-3β,24β-diol) has been found in low levels in human (36,67–73), equine (68,74,75), bovine (76), rat (77), and rabbit brain (73). Dhar *et al.* (73) have clearly established that this compound is not an artifact of isolation techniques by demonstrating its biosynthesis from incubation of [1,2-^{3}H]- or [4-^{14}C]cholesterol with the 105,000g microsomal pellet from bovine cerebral cortical homogenates, and from rat brain homogenates and microsomal fractions. Lin and Smith (78) also followed the uptake into subcellular fractions of [24-^{3}H]cerebrosterol and its 24-epimer [24-³H]24-epicerebrosterol) injected intracerebrally into 18-day old rats. Sacrifice after varying time periods suggested that the removal of [24-³H]cerebrosterol from brain results from an enzymatic metabolism of the sterol, and it was concluded that cerebrosterol exists in a dynamic state of biosynthesis and catabolism. It may be pertinent to mention here that human brain 24-hydroxycholesterol is a single epimer, having the absolute configuration 24β-(24S)-hydroxycholesterol (71). A mixture of isomers would provide impetus for describing the compound as an artifact.

Either 26-hydroxycholesterol or 24-hydroxycholesterol might represent compounds having cleavage points for the biosynthesis of "acidic sterols" in the brain, although there is as yet no experimental evidence for this. However, it may be pertinent to mention that 24-hydroxylated C-26- and C-27-steroidal bile acids are known to occur in nature (79).

Cholestanol (II), another minor sterol found in commercial preparations of cholesterol from bovine brain and spinal cord (80), seems an unlikely candidate for side chain fragmentation in brain tissue. 25-Hydroxycholesterol could conceivably be as likely a candidate as 24-or 26-hydroxycholesterol, but has been considered an artifact of the isolation procedures used with brain tissue (36). We mention them together only because of the excessive accumulation of cholestanol (and cholesterol) in many body tissues, including brain, in the disease cerebrotendinous xanthomatosis (CTX) (81 and references therein). CTX is a rare inherited storage disease. In a recent study Setoguchi et al. (81) also found abnormal accumulation of 5β-cholestane-3α,7α,12α,25-tetraol (XIX) and 5β-cholestane-3α,7α,12α,24,25-pentol (XX) in the bile and feces of three humans (Fig. 3).

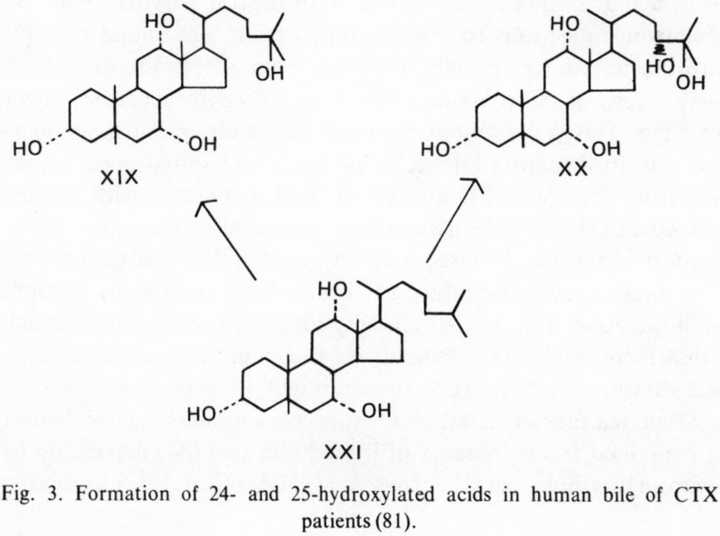

Fig. 3. Formation of 24- and 25-hydroxylated acids in human bile of CTX patients (81).

No 26-hydroxylated intermediates were found. In this case the 25-hydroxylated intermediates were not artifacts. The study suggested that there exists an alternative pathway of cholic acid biosynthesis involving the 25-hydroxylation of 5β-cholestane-3α,7α,12α-triol (XXI). Extrapolating these observations to brain where 25-hydroxycholesterol detection has, as of the present been attributed to artifact formation, makes further comment not pertinent. Based on the above observations made in patients with CTX, 25-hydroxycholesterol and cholestanol metabolism in brain should be investigated wherever possible.

VI. BRAIN, BILE ACIDS, AND THE DEMYELINATING DISEASES

Having presented sufficient data, we trust, to convince the reader that the huge mass of cholesterol in the central nervous system *is not* a metabolically *inert* pool of sterol, we must face the question: What significance do these observations have with respect to the demyelinating diseases? We will present in chronological order the evidence that is currently available. The early suggestion of Weil (1) has been quoted in the Introduction. Probably the first evidence that this conjugated bile acid or cholic acid itself also act as active demyelinating agents *in vivo* came from the observations of Nicholas and Herndon (82) who injected cats intracerebrally with the acids. It is interesting to note that in this study, [^{14}C]carboxyl-labeled sodium cholate injected intracerebrally in considerably high dosages could not be detected in appreciable amounts in central nervous tissue within a week following injection. In a later study (83) it was found that [24-^{14}C]-sodium lithocholate was rapidly released from the brain of guinea pigs, with only traces of ^{14}C remaining two hours after intracerebral injection of 2 μg to 5 mg. This is illustrated in Fig. 4. These observations could account for difficulty in detecting bile acids in brain and spinal cord tissue in the demyelinating diseases, although the wound associated with intracerebral injection could ostensibly be associated with the phenomenon.

A more extensive discussion and intensive effort to detect "steroidal acids" in brain tissue (immature rat brain) was initiated by Nicholas (2). Although the acids were not specifically identified, strong experimental evidence that there might be monohydroxy bile acids present in trace amounts in young brain, where active turnover might be expected, was presented. Brains of guinea pigs afflicted with experimental allergic encephalomyelitis (EAE) contained trace amounts of lithocholic acid (84) detectable by thin-layer chromatography of the free acid and thin-layer chromatography

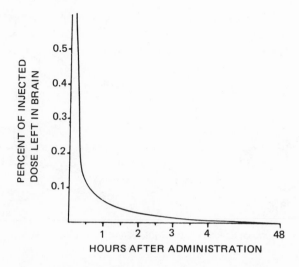

Fig. 4. Plot showing the residual radioactivity in the brain after intracerebral administration of [24-¹⁴C]sodium lithocholate. The doses given were: 2μg (0.1 μci), 1 mg (1 μci), 1.25 mg (1.25 μci), 2.5 mg (2.5 μci) and 5 mg (15 μci). The pattern of excretion was the same with all these doses. [Naqvi et al. (83)].

combined with mass spectrometry of the methyl ester. EAE is an experimental demyelinating disease induced in susceptible species of mammals by subcutaneous injection of brain tissue suspended in an adjuvant and killed tubercle bacillus (85). Some other unidentified monohydroxy bile acids were also indicated. None was detected in normal control guinea pig brain tissue. Subsequent to this, lithocholic acid was detected by the same means (TLC, mass spectrometry) in a coronal specimen of human brain obtained on autopsy of a multiple sclerosis patient (86). The acid was not detected in a somewhat larger coronal specimen of normal human brain. Finally, Naqvi et al. (87) have presented evidence that when incubated with normal guinea pig brain preparations, ³H-labeled (Wilzbach method (88)) 3β-hydroxy-5-cholen-24-oic acid (XXII) is metabolized to lithocholic acid, while 3β-hydroxy-cholest-5-en-26-oic acid (XXIII) and 26-hydroxy-cholesterol similarly labeled are converted to unidentified steroidal acidic products. More significantly, cholesterol so labeled was also converted to steroidal acids (not identified) and this capacity to metabolize cholesterol was enhanced in incubations of brain tissue from EAE-afflicted guinea pigs. Their suggested biosynthetic route from cholesterol to lithocholic acid is shown in Figure 5. Naqvi (89) has also recently shown that one of the acids which remained

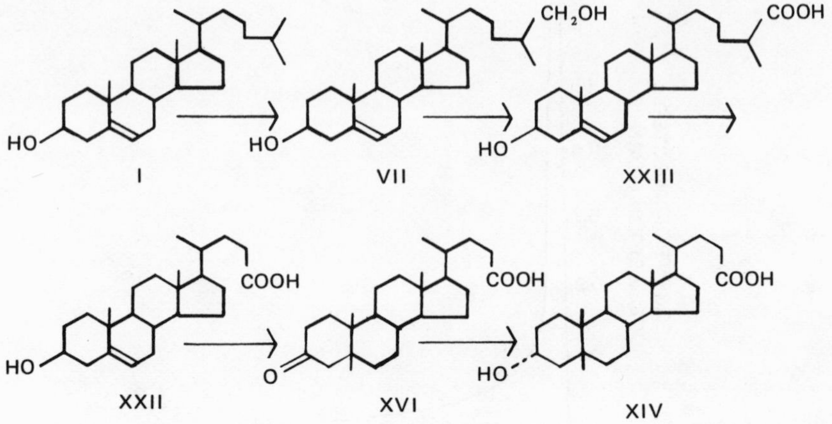

Fig. 5. Possible route for the synthesis of lithocholic acid from cholesterol by brain tissue
[From Naqvi *et al.* (87)].

unidentified in a previous investigation of acidic steroids in EAE tissue (84)
is 3β-hydroxy-5-cholen-24-oic acid.

Palmer (90) has recently summarized the toxic effects of lithocholic
acid and other monohydroxy bile salts in liver disease. Certainly these same
toxic consequences would occur in brain should monohydroxy bile acids
accumulate in abnormal quantity in the central nervous system. More
extensive and conclusive evidence for the latter therefore deserves study.
Perhaps the data presented in this last section may represent the first
experimental evidence for extrahepatic biosynthesis of "bile acids," even
though lithocholic acid is not a normal bile acid in the mammal. Dupont *et
al.* (this volume, Chapter 2) have also directed attention to possible extrahepatic bile acid biosynthesis in the mammal, and their observations (91 and
references therein) should be of considerable future import.

REFERENCES

1. A. Weil, *Arch. Pathol.* **9,** 828 (1930).
2. H. J. Nicholas, *J. Am. Oil Chem. Soc.* **42,** 1008 (1965).
3. P. Morell, M. B. Bornstein, and W. T. Norton, *in* "Basic Neurochemistry" (R. W.
   Albers, G. J. Siegel, R. Katzman, and B. W. Agranoff, eds.), p. 497, Churchill-Living-
   stone, London (1972).
4. "Slow Virus Diseases" (J. Hotchin, ed.) Progress in Medical Virology Vol. 18, Karger,
   Basel (1974).

5. D. McAlpine, C. Lumsden, and E. D. Acheson, "Multiple Sclerosis, A Reappraisal," Churchill-Livingston, London (1972).
6. A. Swann, M. H. Wiley, and M. D. Siperstein, *J. Lipid Res.* **16**, 360 (1975).
7. P. A. Srere, S. S. Chaikoff, S. S. Treitman, and L. S. Burstein, *J. Biol. Chem.* **182**, 629 (1950).
8. J. M. Dietschy and M. D. Siperstein, *J. Lipid Res.* **8**, 97 (1967).
9. J. J. Kabara, *in* "Progress in Brain Research; The Developing Brain" (W. A. Himwich, ed.) Vol. 9, p. 155, Elsevier, N.Y. (1964).
10. R. B. Ramsey and H. J. Nicholas, *in* "Advances in Lipid Research" (R. Paoletti and D. Kritchevsky, eds.) Vol. 10, p. 143, Academic Press, N.Y. (1967).
11. J. J. Kabara, *in* "Progress in Brain Research, Neurobiological Aspects, Maturation and Aging" (D. H. Ford, ed.) Vol. 40, p. 379, Elsevier, N.Y. (1973).
12. R. B. Ramsey, *Biochem. Soc. Trans.,* Vol. 1, p. 341 (1973).
13. D. B. Bowen, A. N. Davison, and R. B. Ramsey, *MTP Internatl. Rev. Sci., Biochem. Series,* Vol. 4, 141 (1974).
14. H. Waelsch, W. M. Sperry, and V. A. Stoyanoff, *J. Biol. Chem.* **135**, 297 (1940).
15. J. J. Kabara and G. T. Okita, *J. Neurochem.* **7**, 298 (1961).
16. M. E. Smith, *in* "Advan. Lipid Res." (R. Paoletti and D. Kritchevsky, eds.) Vol. 5, p. 241, Academic Press, N.Y. (1967).
17. H. W. Moser and M. L. Karnovsky, *J. Biol. Chem.* **234**, 1990 (1959).
18. P. J. McMillan, G. W. Douglas, and R. A. Mortensen, *Proc. Soc. Exp. Biol. Med.* **96**, 738 (1957).
19. F. Chevallier and C. Gautheron, *J. Neurochem.* **16**, 323 (1969).
20. H. J. Nicholas and B. E. Thomas, *Biochim. Biophys. Acta* **36**, 583 (1959).
21. H. J. Nicholas and B. E. Thomas, *J. Neurochem.* **4**, 42 (1959).
22. D. C. DeVivo, M. A. Fishman, and H. C. Agrawal, *Lipids* **8**, 649 (1973).
23. J. Edmond, *J. Biol. Chem.* **249**, 72 (1974).
24. J. F. Weiss, G. Galli, E. Grossi, and P. Paoletti, *J. Neurochem.* **15**, 563 (1968).
25. J. J. Kabara, *Lipids* **8**, 56 (1973).
26. K. Bloch, B. N. Berg, and D. Rittenberg, *J. Biol. Chem.* **149**, 511 (1943).
27. A. N. Davison, J. Dobbing, R. S. Morgan, M. Wajda, and G. P. Wright, Biochemistry of Lipids (G. Popjak, ed.) *Proc. 5th International Conference on Biochem. Problems of Lipids, Vienna,* p. 85 (1958).
28. A. N. Davison, J. Dobbing, R. S. Morgan, and G. P. Wright, *J. Neurochem.* **3**, 89 (1958).
29. D. Kritchevsky and V. Defendi, *J. Neurochem.* **9**, 421 (1962).
30. A. A. Khan and J. Folch-Pi, *J. Neurochem.* **14**, 1099 (1967).
31. M. Spohn and A. N. Davison, *J. Lipid Res.* **13**, 563 (1972).
32. C. Sérougne and F. Chevallier, *Exp. Neurol.* **44**, 1 (1974).
33. C. Sérougne-Gautheron and F. Chevallier, *Biochim. Biophys. Acta* **316**, 244 (1973).
34. R. B. Ramsey and H. J. Nicholas, (Unpublished Observations).
35. J. Edmond and G. Popjak, *J. Biol. Chem.* **249**, 66 (1974).
36. J. E. Van Lier and L. L. Smith, *Texas Rep. Biol. Med.* **27**, 167 (1969).
37. D. Kritchevsky, S. A. Tepper, N. W. DiTullio, and W. L. Holmes, *J. Am. Oil Chem. Soc.* **42**, 1024 (1965).
38. R. Paoletti, R. Fumagalli, and E. Grossi, *J. Am. Oil Chem. Soc.* **42**, 400 (1965).
39. J. McC. Howell, A. N. Davison, and J. Oxberry, *Res. Vet. Sci.* **5**, 376 (1964).
40. J. McC. Howell, A. N. Davison, and J. Oxberry, *Acta Neuropathol.* **12**, 33 (1969).
41. J. N. Cumings, *Brain* **76**, 551 (1953).
42. J. N. Cumings, *Brain* **79**, 554 (1955).

43. C. W. M. Adams, M. Z. M. Ibrahim, and S. Liebowitz, *in* "Neurochemistry" (C. W. M. Adams, ed.), p. 437, Elsevier, N.Y. (1965).
44. R. B. Ramsey, C. W. Martin, L. LaPalio, and H. J. Nicholas, (Unpublished Experiments).
45. W. H. Elliott and P. M. Hyde, *Am. J. Med.* **51**, 568 (1971).
46. E. H. Mosbach and G. Salen, *Am. J. Dig. Dis., New Series,* **19**, 920 (1974).
47. I. Bjorkhem and H. Danielsson, *Mol. Cell. Biochem.* **4**, 79 (1974).
48. B. I. Grosser, *J. Neurochem.* **13**, 475 (1966).
49. B. I. Grosser and E. L. Bliss, *Steroids* **8**, 915 (1966).
50. B. I. Grosser and L. R. Axelrod, *Steroids* **9**, 229 (1967).
51. L. Sholiton, R. T. Marnell, and E. E. Werk, *Steroids* **8**, 265 (1966).
52. P. Knapstein, A. David, W. Chung-Hsiu, D. F. Archer, G. L. Flickinger, and J. C. Touchstone, *Steroids* **11**, 885 (1968).
53. F. Naffolin, K. J. Ryan, I. Davies, V. V. Reddy, F. Flores, Z. Petro, M. Kuhn, R. J. Shite, Y. Takaoka, and L. Wolin, *in* "Recent Progress in Hormone Research" (R. O. Greep, ed.) p. 31, Academic Press, N.Y. (1975).
54. C. W. Martin and H. J. Nicholas, *J. Lipid Res.* **14**, 618 (1973).
55. C. W. Martin and H. J. Nicholas, *Steroids* **19**, 549 (1972).
56. S. H. M. Naqvi and H. J. Nicholas, *Steroids* **16**, 297 (1970).
57. S. H. M. Naqvi and H. J. Nicholas, *Lipids* **8**, 651 (1973).
58. K. Yamasaki, F. Nada, and K. Shimizu, *J. Biochem.* (*Tokyo*) **46**, 739 (1959).
59. H. Danielsson and K. Einarsson, *J. Biol. Chem.* **241**, 1449 (1966).
60. O. Berseus, H. Danielsson, and A. Kallner, *J. Biol. Chem.* **240**, 2396 (1965).
61. C. W. Martin and H. J. Nicholas, *Steroids* **21**, 633 (1973).
62. N. B. Javitt and S. Emerman, *in* "Bile Salt Metabolism" (L. Shiff, ed.) p. 109, Charles C. Thomas, Springfield, Illinois (1969).
63. L. L. Smith and N. L. Pandya, *Atherosclerosis,* **17**, 21 (1973).
64. J. E. Van Lier and L. L. Smith, *Lipids* **6**, 85 (1971).
65. L. L. Smith, J. D. Wells, and N. L. Pandya, *Texas Rep. Biol. Med.* **31**, 37 (1973).
66. A. G. Smith, J. D. Gilbert, W. A. Harland, and C. J. W. Brooks, *Biochem. J.* **139**, 793 (1974).
67. S. DiFrisco, P. DeRuggieri, and A. Ercoli, *Boll. Soc. Ital. Biol. Sper.* **29**, 1351 (1953).
68. A. Ercoli and P. DeRuggieri, *Gass. Chim. Ital.* **83**, 720 (1953).
69. A. Ercoli and P. DeRuggieri, *J. Am. Chem. Soc.* **75**, 3284 (1953).
70. K. Schubert, C. Rose, and M. Burger, *Hoppe-Seyler's Z. Physiol. Chem.* **326**, 235 (1961).
71. J. E. Van Lier and L. L. Smith, *J. Chromatog.* **49**, 555 (1970).
72. L. L. Smith, D. R. Ray, J. A. Moody, J. D. Wells, and J. E. van Lier, *J. Neurochem.* **19**, 899 (1972).
73. A. K. Dhar, J. I. Teng, and L. L. Smith, *J. Neurochem.* **21**, 51 (1973).
74. A. Ercoli, S. DiFrisco, and P. DeRuggieri, *Boll. Soc. Ital. Biol. Sper.* **29**, 494 (1953).
75. L. F. Fieser, W. Y. Huang, and B. K. Bhattacharyya, *J. Org. Chem.* **22**, 1380 (1957).
76. R. Richter and H. Dannenberg, *Hoppe-Seyler's Z. Physiol. Chem.* **350**, 1213 (1969).
77. Y. Y. Lin and L. L. Smith *Biochim. Biophys. Acta* **348**, 189 (1974).
78. Y. Y. Lin and L. L. Smith, *J. Neurochem.* **25**, 659 (1975).
79. G. A. D. Haselwood, *in* "Bile Salts," Methuen, London, p. 17 (1967).
80. L. F. Fieser and M. Fieser, "Steroids," p. 342, Reinhold Publishing Corp. N.Y. (1959).
81. T. Setoguchi, G. Salen, G. S. Tint, and E. H. Mosbach, *J. Clin. Invest.* **53**, 1393 (1974).
82. H. J. Nicholas and B. Herndon, *Neurology* **14**, 549 (1964).
83. S. H. M. Naqvi, B. L. Herndon, L. Del Rosario, and H. J. Nicholas *Lipids* **5**, 964 (1970).

84. S. H. M. Naqvi, B. L. Herndon, M. T. Kelley, V. Bleisch, R. T. Aexel, and H. J. Nicholas, *J. Lipid Res.* **10**, 115 (1969).
85. W. B. Nattews and H. Miller, *in* "Diseases of the Nervous System," p. 261, Blackwell Scientific Publications, Oxford (1972).
86. S. H. M. Naqvi, R. B. Ramsey, and H. J. Nicholas, *Lipids* **5**, 578 (1970).
87. S. H. M. Naqvi and F. Siddiqui, *J. Pakistan Med. Assoc.* **24**, 55 (1974).
88. K. E. Wilzbach, *J. Am. Chem. Soc.* **79**, 1013 (1975).
89. S. H. M. Naqvi, *Lipids* **8**, 766 (1973).
90. R. H. Palmer, *Arch. Internal Med.* **130**, 606 (1972).
91. J. Dupont, S. Y. Oh, and P. Janson, *in* "Bile Acids" (P. P. Nair and D. Kritchevsky, eds.), Vol. 3, p. 17, Plenum Press, New York (1976).

Chapter 2

TISSUE DISTRIBUTION OF BILE ACIDS: METHODOLOGY AND QUANTIFICATION

Jacqueline Dupont, Suk Yon Oh, and Phyllis Janson

Department of Food Science and Nutrition
Colorado State University
Fort Collins, Colorado

I. INTRODUCTION*

The observation of Chaikoff's group (1,2) has been cited as evidence that liver is the only site for cholesterol catabolism to bile acids (3,4). Incubation of various tissue slices of rat liver, kidney, testes, spleen, lung and brain, and beef adrenal, however, resulted in conversion of [26-^{14}C]cholesterol to $^{14}CO_2$, but [4-^{14}C]cholesterol was not oxidized (3). Since liver was found to be much more active than the other tissues, it was concluded that liver was the major site of bile acid synthesis. Harold *et al.* (2) concluded the same from a rat liver perfusion study. Liver has been verified as a major site of bile acid synthesis (5,6) but no direct study has ever been reported to verify the incapability of extrahepatic tissues to transform cholesterol to bile acids.

Grundy and Ahrens (7) have reported the unaccountable loss of ring-labeled cholesterol in their balance studies. Substantial amounts of ^{14}C-radioactivity have been found in the acidic lipid fraction extracted from the nonhepatic tissues of rat (8,9) and miniature swine (10) after [4-^{14}C]-cholesterol was injected into the animals. Since cholesterol is excreted via feces either in the form of other sterols, principally coprostanol, or as acidic

---

* Abbreviations used are as follows: cholic acid, $3\alpha,7\alpha,12\alpha$-trihydroxy-5β-cholanoic acid; taurocholic acid, $3\alpha,7\alpha12\alpha$-trihydroxy-5α-cholan-24-oyl-taurine; TLC, thin layer chromatography; GLC, gas–liquid chromatography; GLC-MS, gas–liquid chromatography mass spectroscopy.

steroids, such substantial ^{14}C-radioactivity recovered in the saponifiable fraction was unexpected and was considered to be most likely due to bile acids.

Recently, Naqvi *et al.* (11) found that guinea pig brain can reduce 3-keto-5β-cholanoic acid, the immediate precursor of lithocholic acid, to lithocholic acid (3-hydroxy-5β-cholanoic acid). Martin and Nicholas (12) showed that lithocholic acid was converted to 3-keto-5β-cholanoic acid in adult rat brain indicating the presence of a cholesterol-3α-hydroxysteroid dehydrogenase enzyme, and that formation of a more polar metabolite, 3α,6β-dihydroxy-5β-cholanoic acid indicated the presence of a 6β-hydroxylating enzyme.

In a study of [4-^{14}C]cholesterol metabolism in our laboratory (9), carbon-14 was recovered in the acidic lipid fraction of the rat carcass, and the proportion of the radioactivity varied from 14–43% in relation to sex, age, and diet. That much ^{14}C activity indicated that its presence could not be a mere artifact of methodology.

These observations led us to search for bile acids in tissues other than liver. The process required testing available methodology and adapting some processes for use with tissues. Extraction and quantitation are described separately, and determinations of tissue concentrations reported.

II. EXTRACTION METHODOLOGY

An article covering different aspects of bile acid analysis appeared in the first volume of this series (13). It covered techniques ranging from extraction to identification and tissue analysis. The aspect of methodology having widest variability and causing most difficulty appeared to be extraction of bile acids from biological material—a problem reviewed at length by Eneroth and Sjövall (13). Some of the pertinent points will be summarized here.

The protonated conjugated bile acids are sufficiently nonpolar to be extracted from aqueous solutions with *n*-butanol (13). It is necessary, however, to have a 10–20 volume excess of the organic solvent compared to the water volume.

The Folch *et al* (14) procedure has been used to obtain a simultaneous extraction and purification, and the conjugated bile acids are then found in the upper (water) phase (15). Manes and Schneider (16) used 20 volumes of 0.5% HCl in 95% ethanol to extract fecal bile acids bound to cholestyramine. The HCl released the resin-bound bile acids as well as protein-bound ones.

All of the liquid–liquid extraction procedures are hampered by the amphipathic nature of the bile acids. In most systems they partition among the phases and are not predominantly concentrated in one. Recently this problem has been eliminated by the introduction of a liquid–solid extraction procedure (17). Schwarz *et al.* (18) have used Amberlite XAD-2 to extract bile acids quantitatively from serum.

III. TESTS OF EXTRACTION METHODS

A. Liquid–Liquid

In quantitative work it is useful to add radiolabeled compounds or internal standards to monitor recovery. For the work reported here [24-^{14}C]lithocholic and cholic, and [26-^{14}C]taurocholic acids were used. [^{14}C]Cholesterol was used to monitor removal of nonsaponifiable material. Purity was checked by thin-layer chromatography. Nonradioactive standards were also purified by TLC. Three liquid–liquid extraction methods and a liquid–solid method have been tested.

The details of our use of the liquid–liquid extraction methods have been reported elsewhere (19). The Folch *et al.* procedure (14), the Manes and Schneider procedure (16), and a generally used saponification procedure with ethanolic KOH were compared. The results are summarized in Table I. The Manes and Schneider procedure is the most satisfactory of liquid–liquid methods for both cholesterol and bile acids. Although it minimizes water in the system, the main drawback of this method is its lack of specificity. Much material which is not cholesterol or bile acids is extracted, and additional isolation procedures are then required.

TABLE I. Recovery of Purified Standards Added to Muscle or Liver Prior to Homogenization by Extraction with Different Methods[a]

Compounds	HCl–EtOH[b]	Alcoholic KOH	Folch[c]
[^{14}C]Taurocholic	82.1%	12.2%	5.5%
[^{14}C]Cholesterol	92.0%	89.9%	89.3%
[^{14}C]Cholic acid	91.6%	81.0%	79.4%
[^{14}C]Lithocholic		88.8%	

[a] Mean of six samples.
[b] Reference (16).
[c] Reference (14).

The KOH digestion should have been more effective for conjugated bile acids than the 12% recovery we obtained. The use of a large excess of solvent, although inconvenient and wasteful, would, however, increase the recovery.

B. Liquid–Solid

The use of Amberlite XAD-2 (Mallinckrodt) for isolation of bile acids from serum was tested by the method reported by Schwarz *et al.* (18). Recovery was monitored by use of standard bile acids checked for purity by TLC. Our results were in the same range as those reported (Fig. 1,Table II). They obtained 82–101% recovery and human serum values of 0.3–9.3 μmol/liter of serum.

The method has been adapted for use with tissues. It is necessary to obtain a homogeneous portion of the tissue to be analyzed and to insure that the bile acids are not bound to other compounds. The method chosen was complete hydrolysis of the tissue with ethanolic KOH. Tissues have a high lipid content, and the digestion process yields fatty acid potassium salts. Details of the procedure for removing fatty acids and recovering bile acids follow.

All reagents were of analytical grade. 3α-Hydroxysteroid dehydrogenase (*Ps. testosteroni*, E 1.1.1.50) (0.5 units/mg) was purchased from Worthington Biochemical Co., Freehold, N.J. Nicotine adenine dinucleotide (NAD) was purchased from Sigma Chemical Co., St. Louis, Mo. Cholic acid was purchased from Steraloids, Inc., Pawling, N.Y.

The glycine buffer containing hydrazine sulfate (Mallinckrodt) and ethylene diamine tetraacetic acid (EDTA, Fisher), the solutions of purified hydroxysteroid dehydrogenase, NAD in buffer, and the working standards containing 0.125, 0.25, 0.5, and 1.0 mmol cholic acid per liter were pre-

TABLE II. Bile Acid Concentration in Portal Blood Serum of Miniature Swine

Sample (μmol/liter)	Recovery of added standard cholic acid (%)
8.2	96
7.4	104
12.6	98.3
5.2	
10.6	

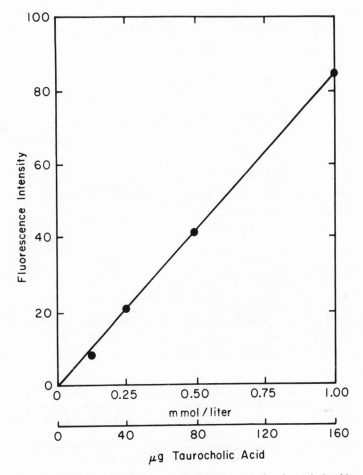

Fig. 1. Typical calibration curve for cholic acid showing relationship between intensity of fluorescence (in galvanometer units) and concentration of standard solution or as related to micrograms taurocholate based upon molecular weight.

pared and stored as indicated by Schwarz *et al.* (18). Columns (pasteur pipettes, approximately 35 × 5 mm) were packed with approximately 0.5 g Amberlite XAD-2 and were individually washed sequentially with five volumes each of water, methanol, acetone, and water, as directed by Makino and Sjövall (17).

Rat tissues were digested by 10% KOH in 50% aqueous ethanol for about 12 hr or until all fat was saponified. When whole carcasses were digested, the tissue contributed the water, and KOH pellets (20% of wt) and one volume of ethanol were added. Half milliliter aliquots were diluted 10-fold with physiological saline and passed over a prewashed column. This was carried out at room temperature at a flow rate of about 0.5 ml/min.

The column was then washed with 5 ml of acidic water (concentrated HCl diluted 1:100 with water), followed by 10 ml of hexane. This step is a modification of the Schwarz *et al.* (18) procedure. Tissues contain excessive lipid compared to bile acids. The acidic water wash converts the potassium salts of fatty acids to free fatty acids without leaving excess acid on the column. The free fatty acids are easily soluble in hexane (20) while the salts are not.

The column was finally eluted with 5 ml of methanol and the methanol eluate evaporated to dryness at room temperature under a stream of air. The residue was dissolved in 3 ml of glycine buffer containing hydrazine sulfate and EDTA. From this, aliquots of 1 ml were taken for the enzymatic assay. All extractions were done in duplicate. The enzymatic assay was done exactly as directed by Schwarz *et al.* (18) using a Hitachi–Perkin Elmer fluorescence spectrometer, Model MPF-2A, except that the samples were not centrifuged for 5 min at 14,000*g*.

Recovery experiments were performed by diluting the standards with 20 ml physiological saline and processing through columns. Known amounts of cholic acid were added to tissue hydrolysate before processing. Recovery of standard cholic acid is shown in Table III. Recovery varied from 89.9–99.4% with a mean of 96.8 ± 1.70.

TABLE III. Recovery of Standard Cholic Acid Alone or Added to Rat Carcass Hydrolysate, Separation by Amberlite XAD-2 Column, and Assay by Enzymic Method[a]

Cholic acid added	Cholic acid recovered	
µg	µg	%
8.2	6.97	85.0
4.1	3.88	94.7
2.05	2.04	99.4
1.025	0.97	94.8
8.2	8.11	98.8
4.1	3.68	89.8
4.1	3.75	91.4
4.1	4.26	104.0
4.1	3.97	96.8
2.15	1.97	96.1
	Mean + SEM	96.8 + 1.70

[a] Reference (18).

The liquid–solid extraction method is much more satisfactory than any of the liquid–liquid systems which we have tried. It is complete and gives a sample of bile acids which is sufficiently pure to be analyzed by the very specific and sensitive enzymatic procedure. The fatty acids are quantitatively removed by the hexane wash, so the procedure may be used for both fatty acid and bile acid purification and subsequent quantitation.

IV. QUANTIFICATION AND IDENTIFICATION METHODOLOGY

It was necessary to verify the identity as well as quantify the acidic material in tissues. Methodology for these procedures was reviewed by Eneroth and Sjövall (13). That review dealt extensively with TLC, GLC, and GLC–MS procedures.

Other quantitation methods for bile acids are based on spectrophotometric (21–22) or fluorospectrophotometric (23) methods. The major problem with use of colorimetric methods in work with biological material is that of specificity. Thus, TLC has been used in quantitative bile acid analysis for purification of bile acid samples before the fluorometric readings (24).

The newest method for quantitation is to use 3α-hydroxysteroid dehydrogenase to reduce NAD to NADH and assay the NADH fluorimetrically. This process has been tested by Schwarz et al. (18), and is specific for 3α-hydroxysteroids. The enzyme does not react with cholesterol or other β-hydroxy compounds.

V. TISSUE CHOLANOIC ACIDS: RESULTS OF QUANTITATION

To separate and quantitate bile acids by GLC they must be deconjugated and derivatized. We used trifluoroacetates of methyl esters and an OV-210 column. In our laboratory the liquid–solid purification of bile acids was developed following the work in which GLC and GLC-MS quantitation occurred, so the more difficult and less satisfactory procedures of ethanolic HCl extraction, deconjugation, and TLC purification were used. These have been described in detail (19). The results obtained are shown in Table IV.

TABLE IV. Approximate Quantities of Total
Cholanoic Acids in Hepatic and Nonhepatic
Tissues of the Rat Determined by GLC[a]

Tissue	μg/g tissue
Liver	158 ± 47[b]
Skeletal muscle	8.1 ± 3.1
Adipose tissue	6.3 ± 1.4
Kidney	25.5 ± 1.6
Pancreas	34.3
Brain	18.5

[a] Reference (19).
[b] Mean ± standard error of 2–3 rats.

The details of our use of GLC-MS methodology have been comprehensively reported, also (19). All the tissue samples which we succeeded in purifying sufficiently for GLC-MS characterization contained compounds corresponding to the mass spectra of standard bile acids and of those of liver. The significant M/E peaks we obtained are shown in Table V.

We have compared the spectrofluorometric method of Levin *et al.* (23) to GLC quantitation using samples of jejunal contents from a miniature swine. The results are shown in Table VI. The spectrofluorometric method gave results 20–100% higher than the GLC method. This may be a combination of overestimation by the fluorometer because of colored contaminants and possibly underestimation by GLC due to the many steps required in the process.

The results of analysis of rat carcasses by the combination of liquid–solid isolation and enzymic assay are shown in Table VII. Rat carcasses (minus head, heart, liver, lungs, gastrointestinal tract, and blood) were from 9-month-old animals fed semipurified diets (25). The total cholanoic acids per carcass were calculated using the molecular weight of taurocholate, since the enzyme assay gives moles of NAD $\rightarrow$ NADH, therefore, moles of cholanoic acid. Previous work suggests that only conjugated bile acids are present. The results indicate 9–59 mg cholanoic acids per carcass. The males had significantly higher (p < .001) total amounts and concentration per gram body weight than females. If the data obtained by GLC are used to calculate an approximate quantity in muscle, liver, kidney, pancreas, brain, and adipose tissue, the result is about 3 mg. The rats used in that study weighed an average of 274 g. They were 5-month-old females. The lower values obtained by enzymic assay of females weighing an average of 311 g are in the same range of magnitude as those calculated from GLC. It is likely that the GLC method underestimated muscle and adipose tissue concentration because of the difficulty of separating fatty acids from bile acids completely by TLC. This is avoided by use of the liquid–solid extraction method.

In these assays cholanoic acids appear to be present in about 4% of the concentration of cholesterol. Liver concentration is higher than carcass concentration in both, but the ratio is about 5:100 cholanoic acid:cholesterol in the liver.

Results of enzymic analysis of rat brains are given in Table VIII. Cholesterol concentration is given also and the ratio of cholanoic acid to cholesterol is about 4:100. These values are more than 10 times those found

TABLE V. Comparison of Major Peaks of Mass Spectra of Standards and Tissue Cholanoic Acids to Those of Sjövall[a]

	Sjövall							
M/E	Di-OH	Tri-OH	GC fragments	Our standard (mono + di-OH)	Liver	Lung	Muscle	Kidney
		486						
484	X			X	X		X	
372	X	X	X	X	X			
370	X			X	X	X	X	
369	X	X		X	X	X	X	X
367	X	X		X	X	X	X	X
342	X	X		X				
329	X	X		X	X		X	
327	X	X		X	X	X	X	
316	X			X	X	X	X	X
276	X		X	X	X			X
257	X	X	X	X	X	X		X
255	X	X	X	X	X	X	X	X
249	X	X	X	X	X			
231		X	X	X	X			X
215	X	X	X	X	X	X	X	X
213	X	X	X	X	X	X	X	X
157	X	X		X	X	X	X	X
154	X	X	X	X	X	X	X	X
149	X	X	X	X	X	X	X	X
142	X	X		X	X	X	X	X
129	X	X	X	X	X	X	X	X
105	X	X	X	X	X	X	X	X
88	X	X	X	X	X	X	X	X

[a] From reference (17).

TABLE VI. Comparison of Total Bile Acid Quantitation by Gas–Liquid Chromatographic and Fluorometric Methods

Sample[a]	GLC (mg)[b]	FL (mg)[c]
P 105	0.442	0.743
P 186	1.374	1.545
P 199	0.399	0.569
P 203	0.760	1.417
P 204	1.082	1.631
P 247	2.510	3.594
P 252	1.732	2.352

[a] All samples are from 200 mg of dry jejunal contents from miniature swine.
[b] See reference (19) for methods.
[c] Reference (23). Although these samples were decolorized by charcoal, some pigment remained and probably caused erroneously high readings.

TABLE VII. Total Cholanoic Acid Concentration Based upon Taurocholate Standard of Eviscerated Rat Carcasses

		Cholanoic acids	
		mg	μg/g body weight
Males			
1		13.18	38.99
2		25.03	38.81
3		15.79	29.35
4		27.60	39.83
5		20.02	31.28
6		21.10	48.51
7		28.17	46.56
8		32.66	59.06
	Mean ± SEM	22. 94 ± 2.34[a]	41.55 ± 3.41[a]
Females			
1		10.57	33.03
2		11.73	37.48
3		9.21	30.40
4		9.44	32.00
5		9.67	30.03
	Mean ± SEM	10.12 ± 0.46	32.59 ± 1.33

[a] Males were higher than females ($p < .001$).

TABLE VIII. Cholesterol and Cholanoic Acid
Concentration of Rat Brain—Determined by KOH
Digestion, Liquid–Solid Extraction, and Enzymic
Assay

Rat	Cholesterol (mg/g brain)	Cholanoic acids (mg/g brain)
20	6.30	0.230
21	6.44	0.204
22	6.12	0.232
23	5.73	0.247

in brains of similar rats by TLC–GLC preparation. The large excess of lipid present in brain tissue may have caused poor recovery of the cholanoic acids.

In summary, all tissues we have examined contain material which we conclude to be cholanoic acids. The evidence suggests that there are several varieties and that they are conjugated. Further description, origin, and fate of the tissue cholanoic acids are under investigation.

REFERENCES

1. J. R. Meier, M. D. Siperstein, and I. L. Chaikoff, *J. Biol. Chem.* **198**, 105 (1952).
2. F. M. Harold, J. M. Felts, and I. L. Chaikoff, *Am. J. Physiol.* **183**, 459 (1955).
3. H. VanBelle, *in* "Cholesterol, Bile Acids and Atherosclerosis," N. Holland Publ. Amsterdam, Holland p. 74 (1965).
4. G. S. Boyd and I. W. Percy-Robb. *Am. J. Med.* **51**, 580 (1971).
5. D. Mendelsohn, L. Mendelsohn, and E. Staple, *Biochemistry* **5**, 3194 (1966).
6. K. A. Mitropoulos and N. B. Myant, *Biochem. J.* **103**, 472 (1967).
7. S. M. Grundy and E. H. Ahrens, Jr., *J. Clin. Invest.* **45**, 1503 (1966).
8. J. Dupont, K. S. Atkinson, and L. Smith, *Steroids* **10**, 1 (1967).
9. J. Dupont, M. M. Mathias, and N. B. Cabacungan, *Lipids* **7**, 576 (1972).
10. J. Dupont, S. Y. Oh, L. A. O'Deen, and S. Geller, *Lipids* **9**, 294 (1974).
11. S. H. M. Naqvi, B. L. Herndon, M. T. Kelley, V. Bleish, and R. T. Aexel, *J. Lipid Res.* **10**, 115 (1969).
12. C. W. Martin and H. J. Nicholas, *Steroids* **19**, 549 (1972).
13. P. Eneroth and J. Sjövall, *in* "The Bile Acids" (P. P. Nair and D. Kritchevsky eds.), Vol. 1 p. 121, Plenum Press New York (1971).
14. J. Folch, M. Lees, and G. H. S. Stanley, *J. Biol. Chem.* **226**, 497 (1957).
15. R. Shioda, P. D. S. Wood, and L. W. Kinsell, *J. Lipid Res.* **10**, 546 (1969).
16. J. D. Manes and D. L. Schneider, *J. Lipid Res.* **12**, 376 (1971).

17. J. Makino and J. Sjövall, *Anal. Lett.* **5**, 341 (1972).
18. H. P. Schwarz, K. V. Bergmann, and G. Paumgartner. *Clin. Chim. Acta* **50**, 197 (1974).
19. S. Y. Oh and J. Dupont, *Lipids* **10**, 340 (1975).
20. P. Schmid and E. Hunter, *Physl. Chem.* **3**, 98 (1971).
21. B. A. Kottke, J. Wollenweber, and C. A. Owen, *J. Chromatog.* **21**, 439 (1966).
22. J. A. Tung and R. Ostwald, *Lipids* **4**, 216 (1969).
23. S. J. Levin, J. L. Irvin, and C. G. Johnston, *Anal. Chem.* **33**, 856 (1961).
24. G. Senemuk and W. Beher, *J. Chromatog.* **21**, 27 (1966).
25. J. Dupont, A. A. Spindler, and M. M. Mathias, *Lipids* (in press).

Chapter 3

BILE-SALT–PROTEIN INTERACTIONS

P. P. Nair

Biochemistry Research Division
Department of Medicine
Sinai Hospital of Baltimore, Inc.
Baltimore, Maryland

I. INTRODUCTION*

The naturally occurring bile acids and their salts form a unique group of biological detergents, exhibiting a spectrum of characteristic physico-chemical properties (1). Unlike the aliphatic anionic detergents, they possess a rigid cyclic ring structure common to all steroids, are amphiphilic in nature and have distinct lipophilic and hydrophilic centers. They are considered as derivatives of cholanoic acid, a C-24 sterol acid in which the ring structure is hydroxylated usually in positions C-3, C-7, and C-12, giving rise to a series of mono-, di-, and trihydroxycholanoic acids (for a review of the chemistry of bile acids, see Chapter 1 in Vol. 1 of this series) (2). While the hydroxyl substituents and the carboxyl group together confer hydrophilic properties to the molecule, the large bulk of the cyclic hydrocarbon retains its lipophilic character. Although the sodium and potassium salts of deoxycholic and cholic acids (di- and trihydroxycholanoic acids) have been in general use as detergents for the solubilization of enzymes from membranes and other particulate preparations, we have been slow in recognizing the importance of more specific interactions between various bile salts and proteins in modulating the biological activity of various macromolecules. There is an increasing body of evidence supporting the idea that bile salts and certain proteins interact through molecular

* The studies reported in this chapter were supported in part by Grant AM-02131 from the National Institutes of Health.

association to bring about specific conformational alterations. In this chapter, I will review the current state of our knowledge in this area and survey its implications in terms of the pathophysiology of certain clinical states. Citations to the literature will be selective, since no attempt will be made to review studies in which bile salts have been used primarily as detergents for the solubilization of membrane-bound enzymes.

II. NATURE OF MOLECULAR INTERACTIONS

When we consider the consequences of binding of small molecules to macromolecules such as proteins, the determinants of structure and conformation are vital to our understanding of biological activity. For instance, the association of bile salts and proteins could involve either covalent or noncovalent interactions. Recent studies conducted in our laboratories have revealed the existence of lithocholic acid, covalently bound to tissue protein (3), and the propensity of this bile acid to spontaneously couple with bovine serum albumin *in vitro* (4). In this instance, unlike other bile acids, lithocholic acid under certain *in vivo* and *in vitro* conditions appears to form covalent bonds, an interaction in which the primary structure of the protein is altered.

Among the noncovalent bonding interactions, there are two types of weak bonds or interaction forces (5). Each of these two types confer either geometrical specificity of three-dimensional conformation or thermodynamic stability on confirmation. The first of these consists of hydrogen bonds and ionic bonds. Positively charged NH_3^+ groups on the side chains of basic amino acids in proteins, while repelling other positively charged groups, will attract negatively charged ones such as the COO^- in bile salts. The existence of such electrostatic bonds between the positively charged ϵ-amino group of lysine and the negatively charged carboxylate group of bile acids has been postulated by Rudman and Kendall (6). Since the strength of this ionic bond is inversely proportional to the square of the distance between the oppositely charged groups, observations on such interactions may yield valuable information on the accessibility of charged sites on proteins and the three-dimensional structure of the protein in the vicinity of the charged group. Ionizable groups that are buried or partially buried in the native protein may not show any affinity for the carboxylate ion of certain bile acids since the distance between the oppositely charged groups might be too large for any interaction. Unmasking of such groups by conformational alterations can be expected to be accompanied by the expression of ionic interactions. Similarly, hydroxy-substituted bile salts could interact through hydrogen bonding, a phenomenon that could be exploited in a carefully defined system.

Hydrophobic interactions (7) and Van der Waal's forces confer thermodynamic stability on the conformation of macromolecules. Hydrophobic interactions in aqueous systems are more important quantitatively than Van der Waal's forces and, hence, the more lipophilic bile acids could be expected to exhibit a significant affinity for hydrophobic centers in proteins. This type of ligand binding could lead to significant alterations to the quarternary structure or conformational state of a protein. In subsequent discussions on the differences in behavior of different bile acids, we will discuss how hydrophobic interactions could have led to differences in their behavior. Furthermore, hydrophobic bonding, an entropy-driven interaction, is assumed to be a dominant mechanism in the binding of small molecules by biological macromolecules (7).

III. COVALENT LINKAGE OF BILE ACIDS TO PROTEINS

Covalently Linked Tissue Bile Acids. The classical technique of extracting bile acids from tissues involves the solubilization of bile acids by homogenization of tissue in a suitable solvent, such as dilute ethanol or 95% ethanol containing 0.1% of ammonium hydroxide (w/v) (8). During the course of our studies on the bile acid composition of liver biopsies from subjects who had undergone the intestinal bypass procedure for obesity, we noticed the difficulty in extracting lithocholic acid from certain samples. This observation led us to suspect the existence of strong interactions of the type found in covalent bonds.

Upon homogenizing tissues (liver) with 95% ethanol/0.1% NH_4OH, almost all of the bile acids are normally extractable into the solvent (8). Dissolution of the residue with base and subsequent enzymatic hydrolysis of the tissue-bound bile acids with cholylglycine hydrolase (9,10) released additional amounts of lithocholic acid in certain tissue samples. Examination of human hepatic tissues at random showed varying amounts of bound lithocholic acid. Table I illustrates the values obtained for soluble and bound lithocholic acid in a series of hepatic tissues. Since cholylglycine hydrolase catalyzes the enzymatic cleavage of the C–N bond in bile acid conjugates, the release of lithocholic acid from tissues is assumed to involve the cleavage of a covalent bond between the bile acid and the basic side groups of proteins. That this enzymatic reaction is capable of releasing bile acids conjugated to macromolecules was shown by the following experiment. [^{14}C]Lithocholic acid was conjugated to bovine serum albumin using

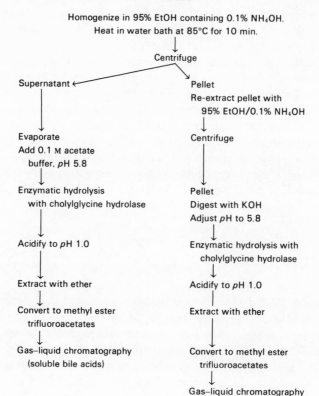

TISSUE

Homogenize in 95% EtOH containing 0.1% NH₄OH.
Heat in water bath at 85°C for 10 min.

Centrifuge

Supernatant

Pellet
Re-extract pellet with
   95% EtOH/0.1% NH₄OH

Evaporate
Add 0.1 M acetate
   buffer, pH 5.8

Centrifuge

Enzymatic hydrolysis
   with cholylglycine hydrolase

Pellet
Digest with KOH
Adjust pH to 5.8

Acidify to pH 1.0

Enzymatic hydrolysis with
   cholylglycine hydrolase

Extract with ether

Acidify to pH 1.0

Convert to methyl ester
   trifluoroacetates

Extract with ether

Gas–liquid chromatography
   (soluble bile acids)

Convert to methyl ester
   trifluoroacetates

Gas–liquid chromatography
   (bound bile acids)

SCHEME I. Outline of procedure for the extraction and quantitative
determination of soluble and bound tissue bile acids.

the water-soluble carbodiimide, 1-cyclohexyl-3-(2-morpholinoethyl) carbodiimide (11). The purified product, added to aliquots of hepatic residues
formed after extraction of tissues with 95% ethanol/0.1% NH₄OH, was
allowed to be hydrolyzed with cholylglycine hydrolase in the presence of all
the necessary cofactors. Approximately 85–90% of the added radioactivity
was recovered in the form of free lithocholic acid which was confirmed by
thin-layer chromatography.

Acid hydrolysis of tissue protein (6 N HCl, 105°C, 18 h in partial
vacuum), followed by thin-layer chromatography revealed a band that was
positive to both steroid and amino acid reagents and showed the same
migration as synthetic ε-lithocholyl lysine. Since synthetic ε-lithocholyl

lysine is almost totally refractory to the acid hydrolysis procedure, it must be concluded that ϵ-lithocholyl lysine is the dominant molecular form in which the steroid appears in tissue proteins.

Nature of the Spontaneous Interaction between Lithocholic Acid and Bovine Serum Albumin. In the course of studies on the modification of free NH_2 groups on bovine serum albumin (BSA) where the basic side chains are coupled to various bile acids through an amide bond using a water-soluble carbodiimide as the coupling agent, it was discovered that lithocholic acid showed spontaneous amino group modifying activity even in the absence of a coupling agent. Upon incubating lithocholic acid with BSA, in the absence of any carbodiimide, there was a progressive decrease in the color reaction with trinitrobenzene sulfonic acid (TNBS), indicating a decrease in the number of free amino groups on BSA. Since masking of amino groups by ionic interaction is overcome by the conditions under which the TNBS reaction is carried out, we considered the possibility of a spontaneous formation of a covalent link. When [24-^{14}C]lithocholate was allowed to react with BSA, the product after purification and acid hydrolysis (6 N HCl, 105°C, 18 h), followed by TLC, showed radioactivity associated with ϵ-lithocholyl lysine (3). We have to assume that under the *in vitro* conditions of our study some amount of lithocholate had formed a peptide link with the ϵ-NH_2 groups on the side chains of lysine. This phenomenon, unique to this bile salt, has not been demonstrated with other bile acids in our studies.

Rudman and Kendall (6), in their studies on the binding of bile salts

TABLE I. Distribution of Lithocholic Acid in Liver Tissue

Tissue number	Lithocholic acid	
	Soluble (μg/g tissue)	Bound (μg/g tissue)
72-M-193	17.6	29.7
72-M-161	22.6	7.6
72-M-158	Trace	29.3
72-M-159	20.5	116.0
72-M-163	43.7	9.1
72-M-220	39.8	48.9
A-24-74S	77.8	4.8
A-18-74S	175.8	10.6
72-M-144	86.1	24.6
72-A-52	Trace	11.2
72-A-46	10.4	4.4
72-A-157	Trace	10.9

and their derivatives to plasma protein fractions, have shown by dialysis–equilibrium experiments the unusually high affinity of the monohydroxy bile salts for serum albumin. Although they were concerned only with electrostatic interactions, they nevertheless showed that binding decreases as the number of hydroxyl groups on the ring system is increased and that bile salts with a single hydroxyl group had the greatest affinity for albumin. The spontaneous, high affinity binding of lithocholate to BSA may have some relationship to its known toxicity, usually seen as an inflammatory response in tissue (12). The importance of studying the physical state and binding of bile salts to macromolecules in disease is shown by the observation that bile salts in the serum of patients with hepatic disease is nondialyzable (13).

IV. MOLECULAR INTERACTIONS AND ENZYMATIC ACTIVITY

The consequences of molecular interactions between specific bile salts and proteins are best exemplified by enzymes that show an absolute requirement for bile salts. It appears to be related to the maintenance of confirmational stability of the macromolecule, and is not entirely due to the detergent properties of the bile salt. In some instances, the highly purified enzyme exhibits an absolute requirement for specific bile salts in order to maintain activity, subunit structure, and stability toward proteolytic inactivation.

A. Sterol Ester Hydrolases

Among various classes of enzymes, the hydrolases form a large group of catalytically active proteins that show specific requirements for bile salts.
Pancreatic Cholesterol Esterase. Pancreatic sterol esterase (sterol ester hydrolases, EC 3.1.1.13) which catalyzes the *in vitro* synthesis and hydrolysis of cholesterol esters requires bile salts for activity (14–17). In both esterification and hydrolysis experiments (either esterification of cholesterol or hydrolysis of cholesterol oleate) Vahouny and coworkers (18,19) showed that although the dihydroxy bile acids (chenodeoxycholic, deoxycholic, and glycodeoxycholic) are as effective as the trihydroxy bile acids (cholic, glycocholic, and taurocholic) in solubilizing the substrates, only the latter group is effective in stimulating enzymatic activity. In hydrolysis experiments, cholesterol oleate is effectively solubilized in mixed micelles of glycodeoxycholate and phospholipids, but enzymatic hydrolysis of the

substrate occurs only after the addition of taurocholate. Since microtitration of the fatty acids and TLC of the enzyme digest showed no complex formation between bile salts and substrate, the evidence is compatible with the idea that cholic acid and its conjugates function as a cofactor for pancreatic juice cholesterol esterase. Furthermore, incubation of rat pancreatic juice cholesterol esterase in the presence of taurocholate protected the enzyme against inactivation by added trypsin. When different bile salts are compared for their relative abilities to counteract tryptic inactivation of a highly purified preparation of rat pancreatic juice cholesterol esterase, taurocholate and taurodeoxycholate afforded complete protection, while $3\alpha,7\beta,12\alpha$-trihydroxycholanoic acid (the 7β-hydroxy analog of cholic acid) was only moderately active (20). Since, in these experiments, taurolithocholate (3α-hydroxy) and taurochenodeoxycholate ($3\alpha,7\alpha$-dihydroxy) were significantly less active than taurocholate ($3\alpha,7\alpha,12\alpha$-trihydroxy) and taurodeoxycholate ($3\alpha,12\alpha$-dihydroxy), the results suggest that the 7α-hydroxy substituent is not an obligatory requirement in the interaction between bile salt and protein.

The apparent molecular weight of purified pancreatic juice cholesterol esterase as determined by sodium dodecyl sulfate (SDS) polyacrylamide gel electrophoresis is 69,000 (21). Gel filtration of this enzyme on Sephadex G200 in the presence of bile salt yields a polymer with a molecular weight of 400,000. The molecular interaction between bile salt and enzyme results in the aggregation of the protomer to give a polymer consisting of six subunits associated with three molar equivalents of the bile salt. The polymerization of the enzyme is reversible as shown by the removal of bile salt by Sephadex gel filtration. Polymerization of the enzyme induced by bile acids other than taurocholate (e.g., taurochenodeoxycholate and taurodeoxycholate) results in polymers that appear to have differences in their conformational characteristics and behavior towards tryptic digestion. These observations show that the bile salt–enzyme interaction exhibits a high degree of structural specificity for the bile salt.

More recent studies by Calame et al. (22) show that after the binding of one molecule of bile acid, the solubilized subunit undergoes a conformational alteration making available additional binding sites for the bile salt. There is a sharp rise in the binding isotherm after approximately one molecule of bile salt is bound per subunit, indicating a cooperative binding-induced conformational change. It is assumed that the driving force for this change involves an interaction between the nonpolar portion of cholic acid and a hydrophobic domain in the enzyme (23). The development of resistance to proteolysis by trypsin and chymotrypsin shows that in the polymerized enzyme, cleavage points bearing lysine and arginine residues have become inaccessible to the proteases.

Other Cholesterol Esterases. In a study of the hydrolysis of cholesterol oleate by rabbit aortic cholesterol esterase, Kritchevsky *et al.*
(24) showed that compared to sodium glycocholate, sodium taurocholate
enhances the activity of the aortic enzyme. Similarly, the enzymes catalyzing the synthesis and hydrolysis of cholesteryl esters in normal aorta of
male rats and rabbits, showed a requirement for sodium taurocholate,
although at higher concentrations the bile salt had an inhibitory action (25).

An acid cholesterol ester hydrolase present in acetone–butanol extracts
from porcine and human aortas, showing a pH optimum between 4.2–4.5,
was activated by the addition of sodium taurocholate (26). Maximum
hydrolysis of [4-^{14}C]cholesterol oleate by this lysosomal enzyme occurred
between 3–4 mM of bile salt, while higher concentrations were inhibitory.
When various bile acids were compared for their ability to promote the
enzymatic hydrolysis of [4-^{14}C]cholesterol palmitate, glycocholate was only
half as effective as the corresponding taurine conjugate, while the taurine
and glycine conjugates of the dihydroxy bile acids, deoxycholic and
chenodeoxycholic, were inactive. Preincubation of the porcine enzyme with
taurocholate was shown to have a protective effect against inactivation by
N-ethylmaleimide and iodoacetate.

The activation of cholesterol esterase by certain bile salts *in vitro* has
physiological significance, as shown by Gallo-Torres *et al.* (27). They
administered both free and conjugated bile salts to rats by stomach tube in
an emulsion containing protein, carbohydrate, monoolein, and [4-^{14}C]cho-
lesterol. Lymph was collected to study the amounts of free and esterified
cholesterol. About 65–70% of lymphatic cholesterol derived from animals
receiving conjugated or unconjugated cholate, was found to be esterified,
while none of this form was detected in lymph from those animals treated
with taurodeoxycholate. The structural specificity exhibited by bile salts in
their interaction with enzymes such as cholesterol esterase have important
implications in pathophysiology of hypercholesterolemia and atherosclerosis. For instance, the composition of biliary bile acids in terms of the
glycine/taurine ratio of the conjugates, the relative abundance of biliary
chenodeoxycholic and cholic acids and the qualitative and quantitative
nature of secondary bile acids generated in the gut, could all influence the
transport and metabolism of cholesterol in the intestines and in the blood.

The existence of a cholesterol ester hydrolase in red cell ghosts was
noticed by Poon and Simon (28). Unlike the pancreatic sterol ester
hydrolase, this enzyme has a lower pH optimum (pH 5.4–5.7) and is
strongly inhibited by sodium taurocholate. Erlanson (29) purified an
esterase from rat pancreatic juice that cleaved a variety of water-soluble
and water-insoluble esters and had a molecular weight of 70,000. Bile salts
strongly activated the esterase, caused the aggregation of the monomer into

an active polymer and at the same time enhanced its ability to withstand proteolytic inactivation. Although this esterase resembles pancreatic cholesterol esterase in many of its characteristics, because of its broad specificity against both lipid and water-soluble substrates the enzyme is classified as a carboxylic esterase. Table II presents a summary of the characteristics of sterol ester hydrolases.

B. Lipases

The action of bile salts as promoters of lipolytic and esterase activities in the intestinal absorption of lipids is well recognized as a specific function for these substances, in addition to their role in stabilizing the intestinal fat emulsion (31). Sodium dehydrocholate (3,7,12-triketocholanoic acid, sodium salt) when administered intravenously to human subjects, reduced serum free fatty acids as well as the activity of postheparin lipoprotein lipase (32). Addition of sodium dehydrocholate to postheparin plasma *in vitro* also reduced lipoprotein lipase activity. Similarly, sodium deoxycholate (33,34), taurocholic and glycocholic acids (35,36) inhibited lipoprotein lipase activity. The intravenous administration of sodium dehydrocholate (10 ml of a 20% solution) counteracts the rise in plasma free fatty acids, induced by the activation of hormone-sensitive lipase with an infusion of 0.5 mg of noradrenalin in 300 ml saline (37).

It is assumed that enzymes and proteins in general absorb to interfaces and are denatured by a process of unfolding. Pancreatic lipase is irreversibly inactivated at a hexadecane–water interface, a process that is prevented by bile salts. Brockerhoff (38) concludes from the results of kinetic studies that bile salts and protein cofactors serve to protect the native structure of lipase and to keep the oil–water interface free of blockage by unfolded (denatured) proteins.

Purified porcine pancreatic lipase loses its activity against bile salt-stabilized long-chain triglyceride substrates when it is passed through DEAE cellulose (39), while the addition of boiled pancreatic extracts restored the activity. The cofactor (colipase) was later isolated and shown to be a heat resistant, low-molecular weight protein (10,000) (40). The cofactor had no effect on lipolysis in the absence of bile salts but was required for optimal lipolysis in the presence of bile salts (41). Rat and human pancreatic lipase that were essentially free of colipase were inhibited by sodium taurocholate and sodium taurochenodeoxycholate in concentrations above their critical micellar concentration, but these bile salt concentrations stimulated lipolysis when colipase was added. These studies show that both colipase and bile salts are required for optimal activation of lipase. Results

TABLE II. Characteristics of Sterol Ester Hydrolases—Interaction with Bile Salts[a]

Enzyme	Substrate	pH optimum	Mol wt	Subunit structure	Action of bile salts	References
Pancreatic cholesterol esterase (hydrolytic and synthetic)	Cholesterol oleate	6.2	70,000 (monomer)	Six	TC, activation, polymerization of monomer-protection against proteolysis TDC and TCDC no activation induces polymerization	(18–22)
Aortic cholesterol esterase	Cholesterol oleate	6.2 (synthetic) 6.6 (hydrolytic)	—	—	TC induces activation GC, mild activation	(24,25)
Acid cholesterol ester hydrolase (aortic)	Cholesterol esters	4.3–4.5	—	—	TC, strong activation GC, weak activation TC, TCDC, GD, and GCDC inactive	(26)
Cholesterol ester hydrolase (red cell membranes)	Cholesterol esters	5.4–5.7	—	—	TC, inhibitory	(28)
Pancreatic carboxyl esterase	Wide substrate specificity, both water soluble and water insoluble esters	6.8	70,000	—	TC, activation polymerization of monomer, protection against proteolysis	(29)
Acyl-cholesterol esterase (adipose tissue)	Cholesterol oleate	6.0	—	—	TC, inhibits enzyme	(30)

[a] TC, taurodeoxycholate; TCDC, taurochenodeoxycholate; GDC, glycodeoxycholate; TDC, taurodeoxycholate; GC, glycocholate; GCDC, glycochenodeoxycholate.

of earlier studies appear to indicate that the forces of binding of cofactor and lipase are enhanced by the presence of bile salts at concentrations above their critical micellar concentration (42). Kimura *et al.* (43) isolated a glycoprotein (mol. wt. 13,000) activator from human pancreatic juice, showing a requirement for Ca^{2+} and deoxycholate for maximal activation of lipase. This glycoprotein activator appears to be different from that of Morgan and Hoffman (41) because the former is not inhibited by sodium taurocholate and sodium taurochenodeoxycholate.

A human pancreatic triacylglycerol lipase acting on tributyrin was purified and characterized by Vandermeers *et al.* (44). This enzyme with approximately 420 amino acid residues and a molecular weight of 46,000 is rapidly inactivated at pH 8.0 when incubated with tributyrin. In this lipase–tributyrin system, colipase and sodium taurodeoxycholate together protected lipase against denaturation. However, by itself, micellar concentrations of bile salt (taurodeoxycholate) inhibited human lipase and also caused a down shift in the optimal pH to 7.5. This is probably attributable to the micellar nature of bile salts. Addition of colipase to this system reversed the inhibition of lipase by taurodeoxycholate and abolished the shift in the optimal pH caused by the bile salt. These authors suggest that one molecule of human lipase combines with one molecule of bovine colipase in the presence of optimum concentrations of sodium taurodeoxycholate (5 mM). This observation is supported by similar findings by Borgström and Erlandson (45). Vandermeers and associates (46) examined rat pancreatic lipase activity in relation to its ability to be adsorbed on emulsified tributyrin in the presence of supramicellar concentrations of sodium taurodeoxycholate, between pH 6.0 and 8.0 and at various colipase concentrations. Colipase, in the presence of taurodeoxycholate, increased the rate of lipase adsorption on its substrate and ensured maximum velocities of the enzymatic reaction. Borgström (47) investigated the interactions between pancreatic lipase and colipase with substrate and bile salts. Kinetic and phase distribution studies of lipase and colipase activities show that lipase binds to hydrophobic interfaces resulting in partial irreversible inactivation. Bile salts at micellar concentrations and pH values above 6.5 displace lipase from its binding to hydrophobic interfaces, resulting in a reversible inactivation. In the presence of bile salts, colipase promotes the binding of lipase to a "supersubstrate" but not to other hydrophobic interfaces and catalytic activity is reestablished.

Human milk has been reported to contain two distinct lipases, one that is activated by serum and inhibited by bile salts, and the other inhibited by serum and activated by bile salts (48). The bile salt stimulated lipase appears to be the principal form of lipase in skim milk, while the serum-stimulated component was present only in the cream. Milk from other species contain only a serum-stimulated lipase as the predominant lipolytic enzyme. The bile salt stimulated enzyme is active against emulsified water-insoluble substrates (trioleylglycerol and tributyrylglycerol) and a water-soluble substrate (p-nitrophenyl acetate) (49). The enzyme is stable in milk and in buffer solutions, but as in the case of the pancreatic enzyme, it is rapidly inactivated in the presence of emulsified triacylglycerols. Bile salts protect the enzyme from such inactivation. This lipase in human milk might have an important physiological function, in view of its absolute dependence on bile salts for the hydrolysis of milk triacylglycerols (50). It is also interesting to point out the structure–activity relationships between the bile salts and the enzyme. The ability to activate the enzyme was limited to the primary bile salts (sodium salts of cholic and chenodeoxycholic acids and their glycine and taurine conjugates), while sodium deoxycholate and its taurine and glycine conjugates were inactive. In addition to its role in the activation of the lipase, bile salts afforded complete protection to the enzyme against inactivation by trypsin and chymotrypsin at pH 6.5, in a manner similar to that observed with pancreatic sterol ester hydrolase.

Since bile salts have been shown to interact with lipases and other hydrolases to bring about specific alterations in the catalytic behavior of these proteins, the very natural question would be, how do bile salts interact with similar enzymes at the cellular level in the intestinal mucosa. In a study on the effects of bile diversion on the reesterification of lipids of the rat small bowel, Tandon et al. (51), showed that rates of reesterification were significantly depressed in bile-depleted experimental animals compared to the controls. Direct enzymatic assays of acyl-coenzyme A ligase and acyl-coenzyme A monoglyceride acyltransferase showed that the activities of these enzymes were greater in the control rats compared to those in the experimental group. Furthermore, upon infusing artificial bile intraduodenally in the experimental animals, the enzymatic activities could be maintained at normal levels, indicating the dependency on biliary constituents. These studies provide indirect evidence that bile salts could regulate the activities of lipid reesterifying enzymes at the level of the intestinal mucosa. In addition to cells from the intestinal mucosa, pancreatic homogenates from young rats show the presence of a bile salt stimulated lipase (52).

C. Esterases

There are several enzymes that exhibit esterase activity against synthetic substrates such as the esters of amino acids. The effect of bile salts on the kinetics of these systems and on the conformation of the macromolecules involved, has been a subject of considerable speculation.

Thrombin, the enzyme that catalyzes the formation of fibrin, and plasmin, the fibrinolytic enzyme, both belong to a class of biologically active proteins involved in the maintenance of normal hemostasis. Curragh and Elmore (53) were the first to show that bile salts enhanced the kinetics of thrombin-catalyzed esterase activity against certain synthetic amino acid esters. Similarly, Engel and Alexander (54) showed that cholate and other bile salts promoted the rate of hydrolysis of tosyl-L-arginine methyl ester (TAME) by thrombin. They also showed that both cholate and deoxycholate elevated the fibrinogen clotting activity of thrombin. Exner and Koppel (55) extended these studies to include the esterase activities of the thrombolytic enzyme, plasmin, against tosyl-L-lysine methyl ester (TLME) and TAME. Among the various bile salts tested in this system, only sodium cholate, and its glycine and taurine conjugates, increased the rate of plasmin-catalyzed hydrolysis of TLME. While deoxycholate showed a transient activation of the enzyme at low concentrations, chenodeoxycholate ($3\alpha,7\alpha$-dihydroxycholanoic acid) and lithocholate (3α-hydroxycholanoic acid) were distinctly inhibitory to the enzyme. In contrast to the high structural specificity exhibited by cholate and its conjugates, these bile salts uniformly enhanced the plasmin-mediated hydrolysis of other N-α-substituted L-lysine and L-arginine methyl esters. The enhancement of esterase activity against TLME, by glycocholate is not limited to plasmin, since both thrombin and trypsin were also activated by the bile salt. From structure–activity relationships, it is inferred that an increase in the hydrophobic character of the bile salt makes it not only inactive but a potent inhibitor of plasmin esterase activity. This observation might have important physiological implications.

Cole (56) studied the kinetics of thrombin-catalyzed hydrolysis of TAME as influenced by cholate and other steroids. Under optimum conditions, cholate not only enhanced the hydrolysis of TAME, but also changed its kinetics to apparent zero-order to complete substrate hydrolysis. The action of cholate in this system could be mediated through two distinct mechanisms, one of which proposes the formation of soluble complexes of TAME and the bile salt. These complexes are visualized as polymeric molecules of TAME and cholate, in which the hydrophobic tosyl groups (directed inwards) interact with the hydrocarbon moiety of the steroid, while the hydrophilic methyl arginyl groups of TAME are oriented towards the surface and into the aqueous medium where the ester bonds are accessible to thrombin. The other mechanism that has been postulated involves direct binding of the steroid to the catalytic protein (thrombin) resulting in conformational changes favorable for enzymatic activity. Cole (56) has proposed a possible interaction between the steroid and the arginyl residues of thrombin although electrostatic interaction with the basic side groups of the arginyl residues is unlikely to occur at pH 7.6, the pH at which the enzyme is assayed.

Although the physiological implications of these findings are not clear, it is known that rat pancreatic juice and serum esterases, both acting on p-nitrophenyl and β-naphthyl acetates are enhanced by conjugated bile salts (57,58).

D. Arylsulfatases

Arylsulfatase A (aryl-sulfate sulfohydrolase) is a lysosomal enzyme that hydrolyses cerebroside sulfate and the synthetic substrate 4-nitrocatechol sulfate. In an assay system developed for cerebroside sulfatase activity in cells from human fibroblast cultures, there was no activity in the absence of sodium taurodeoxycholate (59). Other detergents such as cetyltrimethylammonium bromide, cutscum, Triton X100, Triton WR 1339, Tween 40, Tween 80, and Brij 35 were inactive. Cholate and deoxycholate were about one-half and one-tenth as active as taurodeoxycholate. A similar requirement for taurodeoxycholate has been demonstrated for cerebroside sulfatase from several invertebrates (60), with one exception, the garden snail, *Helix pomatia* in which the bile salt caused an inhibition. Fluharty *et al.* (61) demonstrated an obligatory requirement for cholate or taurodeoxycholate for arylsulfatase-A-mediated hydrolysis of testicular sulfoglycerogalactolipid. Highly purified arylsulfatase A from human urine also required either sodium taurodeoxycholate or sodium cholate for activity towards the physiological substrate, cerebroside sulfate (62). It has been suggested that bile salt micelles are required for solubilization of the hydrophobic substrate, which otherwise forms large aggregates (63), although this concept is not compatible with the fact that other detergents cannot replace bile salt in the incubation mixture. There is also evidence suggesting a direct bile-salt–enzyme interaction leading to the activation of the enzyme (64). Cerebroside sulfatase deficiency has been reported in the genetic sphingolipidosis, metachromatic leukodystrophy.

E. β-Galactosidases and Amylase

Deficiencies in glycosphingolipid β-galactosidases have been implicated in several genetic disorders. In recent years several systems have been developed for the assay of β-galactosidase activity using different substrates. The activity of galactosyl ceramide and lactosyl ceramide β-galactosidase, measured in leukocyte and fibroblast homogenates showed an activation by optimal concentrations of sodium taurocholate (65). Lactosyl ceramide β-galactosidase in human tissues consists of two genetically distinct enzymes according to Tanaka and Suzuki (66). The effect of pure and crude preparations of taurocholate on the two enzymes, is one of the distinguishing criteria that the authors have used for differentiating the two genetically determined enzymes. One of them lactosylceramidase 1 is present in normal human brain, while the other, lactosylceramidase 2 is the predominant form of β-galactosidase in normal liver. Lactosylceramidase 1 is activated by both pure and crude taurocholate while lactosylceramidase 2 is activated only by the crude bile salt. From the evidence presented it appears that only one form of the enzyme (lactosylceramidase 1) is capable of interacting with pure taurocholate to yield a catalytically active protein. This is perhaps an interesting example of how two genetic variants of the same protein exhibit differences in their response to taurocholate. Purified rat brain glycosphingolipid β-galactosidases show pH optima between 3.9–4.2, depending upon the substrate used, and in all instances sodium taurocholate was the most effective activator of the enzyme (67).

α-Amylase is another enzyme similar to β-galactosidase, the catalytic activity of which is modified by bile salts. A stimulation of amylase-mediated hydrolysis of starch by bile and deoxycholate was first reported by Walker and Hope (68). Recent studies have shown that certain proteins and bile salts such as taurodeoxycholate increase α-amylase hydrolysis of insoluble Cibachron Blue Starch (69). Since the effect is more pronounced in the acidic region the phenomenon is of some physiological significance, due to the fact that the pH of the jejunum is between 5.5–6.5. O'Donnell and coworkers (70) studied the effects of various bile acids, both conjugated and unconjugated, on the activity of α-amylase on insoluble Cibachron Blue Starch. With increasing amounts of sodium cholate, the activity reached a plateau at 160% of the nonactivated enzyme, and further increases in cholate concentration had no effect on the activity of the enzyme. With sodium deoxycholate, a maximum activity of 150% of the control was reached at a concentration of 0.75 mmol/liter, and thereafter, further increases in deoxycholate levels resulted in a progressive inhibition. Both chenodeoxycholate and lithocholate inhibited the enzyme in a concentration-dependent manner, up to the limits of their solubility, because free bile acids precipitate from solution at pH 6.0. When these experiments were repeated with conjugated bile salts, only sodium taurocholate and sodium taurochenodeoxycholate activated the enzyme. Taurochenodeoxycholate goes through a maximum rather quickly, after which the bile salt is inhibitory to the enzyme. The inhibition of amylase hydrolysis of insoluble blue starch, could be prevented by the addition of albumin in the reaction mixture.

There is a close resemblance in the bile salt mediated activation of amylase and that of plasmin-catalyzed hydrolysis of tosyl-L-lysine methyl ester (55) in that both of them are activated by trihydroxycholanoic acid (cholate) and its conjugates, while dihydroxy- and monohydroxy- cholanoic acids exhibit varying degrees of inhibition upon these enzymes. In other words, the ability of cholate (trihydroxycholanoic acid) to stimulate these enzymes is lost when it is replaced by bile salts with lesser number of hydroxyl groups (increased hydrophobic character). In the case of taurocholate, activation of amylase begins at concentrations far below its critical micellar concentration, and therefore, it would appear that micelle formation and related physical phenomena are not directly involved in the mechanism of activation. The low K_i values for chenodeoxycholic acid (CDC) inhibition of amylase hydrolysis of both insoluble and soluble starch (0.15 and 0.26 mmol/liter, respectively) shows that this is a potent inhibitor of α-amylase. Since albumin exerts a protective influence on this inhibition, the deleterious effects of CDC may not be manifested under normal physiological conditions. However, the use of CDC as a therapeutic agent for the dissolution of gallstones may be an instance where we could expect some side effects arising from the presence of undigested starch. Since lithocholate is a potent hepatotoxin, and since we have demonstrated a spontaneous binding of lithocholate to bovine serum albumin in the form of peptide bond formation with the ϵ-amino groups of lysine residues (4), the inhibitory action of this bile salt is not totally unexpected. Studies to investigate the nature of the lithocholate–enzyme interaction would be of interest to determine whether covalent bonds of any type are formed during the course of the inhibition.

F. Nucleotidases and Phosphatases

There are several reports in the literature concerning the catalytic behavior of these enzymes in the presence and absence of certain bile salts. Since some of these enzymes are membrane-bound or are in the form of particulate vesicles associated with lipoproteins, the exact role of bile salts in these systems is not clearly understood. They may vary from purely detergent effects to highly specific ligand–macromolecule interactions resulting in conformational alterations. Therefore, the purpose of this section is to make citations to the literature on a selective basis to assist the reader in obtaining a general overview of the field.

The activity of 5'-nucleotidase in partially purified plasma membranes is enhanced by deoxycholate, regardless of whether the bile salt is added during incubation or is preincubated with the enzyme (71-73). Since removal of deoxycholate from enzyme preparations by dialysis did not decrease the elevated levels of enzymatic activity (74) it is assumed that some of the bile salt is firmly bound to the enzyme or that it had induced a conformational alteration favorable for optimal activity. In more recent studies, both deoxycholate (DOC) and Triton X100 have been shown to cause a partial solubilization of the membrane-bound enzyme (75) to yield soluble and particulate fractions. In these preparations, only the particulate enzyme is activated by DOC.

In a study of the effect of various detergents on adenosine $2',3'$-cyclic nucleotide $3'$-phosphohydrolase (CNP), a membrane-bound enzyme associated with the myelin of nervous tissue, Lees *et al.* (76) demonstrated a concentration-dependent rise in enzyme activity when sodium deoxycholate, Triton X100, or cetyltrimethylammonium bromide was added.

The action of bile salts on ATPase is variable depending upon the tissue of origin and the physical state of its aggregation. In the intestinal mucosa, glycocholate at physiological concentrations is reported to be inhibitory to Mg^{2+}- and $Na^+ + K^+$-dependent ATPase activity (77). In contrast to this report, ATPase from mucosal homogenates of rat jejunum and ileum appear to be stimulated by sodium taurocholate and glycocholate in the presence of $Na^+ + K^+$ and Mg^{2+} (78). The differences in the response to bile salts is perhaps attributable to the fact that the earlier workers (77) had used the histidine salt of glycocholate and the 2-amino-2-methyl-1-propanol salt of ATP in their assays. When a purified preparation of sodium and potassium ion-activated ATPase (Na,K-ATPase) from *Squalus acanthias* is incorporated into a synthetic vesicle containing lecithin and cholate, the Na,K-ATPase activity is markedly inhibited by cholate (79). In similar studies on ATPase in phospholipid vesicles, Knowles and Racker (80) isolated a protein from sarcoplasmic reticulum which catalyzed the Ca^{2+}-dependent hydrolysis of ATP. Phospholipid vesicles reconstituted with this protein, required about 1.5 mg of deoxycholate per milligram of protein to keep the enzyme soluble and also to maintain maximal Ca^{2+} transport activity of the preparation.

D-Glucose-6-P phosphohydrolase is another member of the group of phosphohydrolases that exhibit differences in its catalytic activity under the influence of bile salts. When intact rat liver microsomes were exposed to taurocholate (less than 1 mM), the latter inhibited the glucose-6-P phosphohydrolase activity in the membranes (81). Subsequent solubilization of the preparation by either higher concentrations of taurocholate or 1M NH_4OH abolished the inhibition, showing that membrane integrity is essential for the demonstration of taurocholate inhibition. Deoxycholate, on the other hand, activates carbamyl-P: glucose phosphotransferase and glucose-6-P phosphohydrolase from isolated nuclear membranes by 61% and 58%, respectively (82).

The ATPase activity of highly purified rat liver alkaline phosphatase is not stimulated by bile salts when they are added to the incubation mixture (83). However, when the enzyme is preincubated for 60 min at 37°C, taurocholate and taurodeoxycholate increased ATPase activity by 60% and glycocholate by 30%. Under the same conditions, unconjugated bile salts and other detergents had no stimulatory effect. It is assumed that the enzyme undergoes conformational alterations by specific interactions with bile salts during the period of incubation.

G. Peptide Bond Hydrolases

The digestion of protein in the alimentary tract is carried out by proteolytic enzymes derived from precursors known as zymogens, stored in the form of granules in the organ of origin. The stomach and the exocrine pancreas together release zymogens such as pepsinogen, trypsinogen, chymotrypsinogen, and proelastase. The conversion of the zymogens to their active products could be carried out autocatalytically or by the mediation of other activators and proteolytic enzymes such as trypsin. Since gastric and pancreatic secretions come under the influence of bile in the duodenum, the interaction between bile salts and proteolytic enzymes in the activation of their precursors is a subject of considerable significance in the pathophysiology of digestive diseases. Furthermore, through electrostatic and/or hydrophobic interactions, bile salts could presumably accord protection to biologically active peptides against premature proteolytic degradation. Frazer and Schulman (84) reported that bile salts exercise a protective influence on trypsin.

Unlike this observation, in a larger study on the influence of bile salts upon digestive enzymes, Vahouny and Brecher (85) using model substrates did not observe any activation or change in stability of pancreatic juice trypsin. On the other hand, they found that taurocholate when added to pancreatic juice accelerated the tryptic activation of chymotrypsinogen. Hadorn *et al.* (86) and Rutgeerts *et al.* (87) have shown an increase in the velocity of enterokinase-catalyzed activation of trypsinogen. Glycodeoxycholate and taurocholate at a concentration of 2.5 mM caused a 5.8-fold increase in the rate of activation at low trypsinogen concentrations (86). The K_m for the reaction was 0.09 mM in the absence of bile salts and it decreased to 0.015 mM in their presence. Since these authors had used the soluble fraction from rat and human intestinal homogenate as the source of enterokinase, the action of bile salts cannot be ascribed to solubilization of membrane-bound enterokinase. Further physicochemical studies will be needed to elucidate the nature of the interaction and its role in the regulation of proteolysis in the intestines.

Elastase, another pancreatic proteolytic enzyme that attacks insoluble elastin, is activated by sodium dodecyl sulfate as a result of detergent-induced changes to the substrate (88). Among naturally occurring compounds, fatty acids and bile salts enhance the rate of elastolysis when they are added to the assay system (89). Cholate, deoxycholate, glycocholate, and taurodeoxycholate stimulated the rate of hydrolysis about fivefold at concentrations between 15–20 mM. Apparently from circular dichroic spectral data, the authors have concluded that the stimulation of elastolytic activity involves ligand-induced conformational changes to the elastin polypeptide chains. Distinct hydrophobic and ionic requirements dictate the binding of the ligand and rate of catalytic digestion. Compounds such as bile salts which are anionic and have a fairly large hydrophobic region bind to elastin and form tight complexes. A reciprocal charge–charge interaction between enzyme and substrate is essential for elastolysis because maleylation of elastase which renders the enzyme inactive (anionic at pH of assay) for elastin, will hydrolyze elastin if the substrate is rendered cationic by interaction with a cationic ligand.

Folic acid is known to occur in plant and animal tissue in the form of conjugates with polyglutamic acid through a C–N bond between the γ-carboxyl and α-amino group. In mammals these conjugates are presumably first hydrolyzed by a γ-glutamyl carboxypeptidase (folic acid conjugase) prior to their absorption. Since, in certain malabsorption syndromes associated with folic acid deficiency, there is an alteration in bile acid metabolism, Bernstein et al. (90) studied the influence of bile salts on folic acid conjugase. When the enzyme was assayed in phosphate buffer at pH 6.1, cholate did not show any alteration in enzymatic activity up to a concentration of 2.0 μmol/ml, and above that there was a slight inhibition. In contrast, dihydroxycholanoic acids such as deoxycholate and chenodeoxycholate showed marked inhibition of conjugase activity between 0.5–1.0 μmol/ml and 0.05–0.2 μmol/ml, respectively.

Pepsin, which has a pH optimum of 2.0, is markedly inhibited by canine bile at that pH (91). Both taurine and glycine conjugates of cholic acid reduced the activity of pepsin in proportion to their concentration in the assay mixture (92). In comparison to the cholic acid conjugates, glyco-chenodeoxycholic acid taurochenodeoxycholic acids were more potent inhibitors of pepsin.

H. Oxidoreductases

The influence of bile salts on oxidoreductases are generally considered to be the result of a solubilization of membrane-bound enzymes, and, hence, references to the literature will be limited to examples where a specific interaction with the protein is postulated.

Theorell and associates (93) isolated in crystalline form a steroid-active liver alcohol dehydrogenase (LADHS) that oxidizes 3β-hydroxy-5β-cholanoic acid (3β-epimer of lithocholic acid). In this system, lithocholate (3β-hydroxycholanoic acid) at pH 9.5 showed competitive inhibition of the reduction of NAD^+ in the presence of 3β-hydroxy-5β-cholanoic acid, but was practically inactive as an inhibitor when ethanol was used as the substrate. The use of lithocholate in this system as a competitive inhibitor specific for the steroid active site, led to the conclusion that the binding sites for ethanol and for 3β-hydroxy cholanoic acid must be different and independent of each other. These observations have been confirmed in later studies on polymorphic forms of liver alcohol dehydrogenase (94).

Lee and Whitehouse (95) in an extensive study of the inhibition of electron transport enzymes and coupled phosphorylation in hepatic mito-chondria by cholanoic acids, concluded that the most active inhibitors (bile acids) of electron transport were also the most potent uncouplers of oxidative phosphorylation. Ketocholanoic acids and their conjugates were generally less potent than the corresponding hydroxy compounds.

I. Other Bile-Salt–Protein Interactions

Deoxycholate and chenodeoxycholate in the nonionized form interact with receptors in the isolated rat uterus and exert an antioxytocin and antiphosphodiesterase activity (96). Hanson *et al.* (97) isolated a protein from rat liver cytosol, having a molecular weight of 16,000, that binds and solubilizes a series of nine C-27 sterol precursors of bile acids and four different bile acids and their conjugates. This protein appears to be structurally similar to the Sterol Carrier Protein (SCP) required for cholesterol synthesis from its precursors, and it increased the activity of the enzyme 12α-hydroxylase, involved in the biosynthesis of cholic acid. The activation of the enzyme appears to be the result of a specific interaction of a bile-acid-precursor–SCP complex with the active site of the microsomal 12α-hydroxylase.

Accatino and Simon (98) characterized a cholic and taurocholic acid binding receptor on hepatocyte surface membrane, as a liver plasma membrane protein. They suggest that this receptor aids in the transfer of bile acid from plasma into bile.

Makino *et al.* (99) investigated the interaction of deoxycholate and Triton X100 with bovine serum albumin. They showed the existence of four major sites of high affinity binding and about 14 weaker sites for deoxycholate.

V. PATHOPHYSIOLOGICAL IMPLICATIONS

Bile salts as natural constituents of bile could be expected to interact with biologically active proteins in the liver as well as in the gastrointestinal tract. In this review we have seen that there are several factors that determine the mode of bile-salt–protein interactions. The state of activation of enzymes, polypeptide hormones, and cellular receptors are some examples of possible bile salt mediated regulation of biological processes.

The presence or absence of bile salts in the wrong places could also be expected to play a role in the etiology of several gastrointestinal disorders. As an example, interstitial injections of sodium taurocholate and cephalothin into the pancreas induced lethal acute necrotizing pancreatitis in the guinea pig (100). In this instance, taurocholate may have prematurely activated pancreatic zymogens or blocked the activity of endogenous trypsin inhibitor or acted merely as a detergent. More fundamental studies on the nature of bile-salt–protein interactions, particularly with the secondary bile salts, are needed to increase our understanding of biological mechanisms regulated by them.

REFERENCES

1. D. M. Small, *in* "The Bile Acids" (P. P. Nair and D. Kritchevsky, eds.), Vol. 1, p. 249, Plenum Press, New York (1971).
2. D. Kritchevsky and P. P. Nair, *in* "The Bile Acids" (P. P. Nair and D. Kritchevsky, eds.), Vol. 1, p. 1, Plenum Press, New York (1971).
3. P. P. Nair, M. Vocci, J. Bankoski, A. I. Mendeloff, and C. Lilis (in preparation).
4. P. P. Nair, M. Vocci, J. Bankoski, M. Gorelik, and R. Plapinger (Communicated) *Biochemistry*.
5. A. L. Lehninger, *in* "The Neurosciences–A Study Program" (G. C. Quarton, T. Melnechuk, and F. O. Schmitt, eds.), p. 35, The Rockefeller University Press, New York (1967).
6. D. Rudman and F. E. Kendall, *J. Clin. Invest.* **36**, 538 (1957).

7. W. Kauzmann, *Adv. Protein. Chem.* **14**, 1 (1959).
8. T. Okishio, P. P. Nair, and M. Gordon, *Biochem. J.* **102**, 654 (1967).
9. P. P. Nair, M. Gordon, and J. Reback, *J. Biol. Chem.* **242**, 7 (1967).
10. P. P. Nair, *in* "Bile Salt Metabolism" (L. Schiff, J. B. Carey, and J. M. Dietschy, eds.), p. 172, C. C. Thomas, Springfield, Illinois (1969).
11. J. C. Sheehan and J. J. Hlavka, *J. Org. Chem.* **21**, 439 (1956).
12. R. H. Palmer, *in* "Bile Salt Metabolism" (L. Schiff, J. B. Carey, and J. M. Dietschy, eds.), p. 184, C. C. Thomas, Springfield, Illinois (1969).
13. D. Rudman and F. E. Kendall, *J. Clin. Invest.* **36**, 530 (1957).
14. S. V. Nedswedski, *Z. Physiol. Chem.* **236**, 69 (1935).
15. L. Swell, H. Field Jr., and C. R. Treadwell, *Proc. Soc. Exp. Biol. Med.* **84**, 417 (1953).
16. M. Korzenovsky, E. R. Diller, A. C. Marshall, and B. M. Auda, *Biochem. J.* **76**, 238 (1960).
17. S. K. Murty and J. Ganguly, *Biochem. J.* **83**, 460 (1962).
18. G. V. Vahouny, S. Weersing, and C. R. Treadwell, *Biochem. Biophys. Res. Commun.* **15**, 224 (1964).
19. G. V. Vahouny, S. Weersing, and C. R. Treadwell, *Biochim. Biophys. Acta* **98**, 607 (1965).
20. J. Hyun, H. Kothari, E. Herm, J. Mortenson, C. R. Treadwell, and G. V. Vahouny, *J. Biol. Chem.* **244**, 1937 (1969).
21. J. Hyun, C. R. Treadwell, and G. V. Vahouny, *Arch. Biochem. Biophys.* **152**, 233 (1972).
22. K. B. Calame, L. Gallo, E. Cheriathundam, G. V. Vahouny, and C. R. Treadwell, *Arch. Biochem. Biophys.* **168**, 57 (1975).
23. J. Steinhardt and J. Reynolds, *in* "Multiple Equilibria in Protein, Molecular Biology," p. 13, Academic Press, New York (1969).
24. D. Kritchevsky, S. A. Tepper, and G. Rothblat, *Lipids,* **3**, 454 (1968).
25. H. V. Kothari, B. F. Miller, and D. Kritchevsky, *Biochim. Biophys. Acta* **296**, 446 (1973).
26. A. G. Smith, C. J. W. Brooks, and W. A. Harland, *Steroids Lipids Res.* **5**, 150 (1974).
27. H. E. Gallo-Torres, O. N. Miller, and J. G. Hamilton, *Arch. Biochem. Biophys.* **143**, 22 (1971).
28. R. W. Poon and J. B. Simon, *Biochim. Biophys. Acta* **384**, 138 (1975).
29. C. Erlanson, *Scand. J. Gastroenterol.* **10**, 401 (1975).
30. J. Arnaud and J. Boyer, *Biochim. Biophys. Acta* **337**, 165 (1974).
31. P. Desnuelle, *Adv. Enzymol.* **23**, 129 (1961).
32. A. Štork, E. Fabian, and J. Šponarová, *Enzymol. Biol. Clin.* **9**, 137 (1968).
33. H. Engelberg, *J. Appl. Physiol.* **11**, 155 (1957).
34. F. Sandhofer, S. Sailer, and H. Braunsteiner, *Dtsch. Med. Wochenschr.* **89**, 426 (1964).
35. M. F. Crass and H. C. Meng. *Am. J. Physiol.* **206**, 610 (1964).
36. D. Grafnetter and T. Zemplenyi, *Hoppe-Seyler's Z. Physiol. Chem.* **316**, 218 (1959).
37. A. Štork, E. Fabian, and J. Fabianova, *Enzyme* **12**, 269 (1971).
38. H. Brockerhoff, *J. Biol. Chem.* **246**, 5828 (1971).
39. B. Basky, E. Klein, and W. F. Lever, *Arch. Biochem. Biophys.* **102**, 201 (1963).
40. M. F. Maylié, M. Charles, C. Gache, and P. Desnuelle, *Biochim. Biophys. Acta* **229**, 286 (1971).
41. R. G. H. Morgan and N. E. Hoffman, *Biochim. Biophys. Acta* **248**, 143 (1971).
42. R. G. H. Morgan, J. Barrowman, H. Filipek-Wender, and B. Borgström, *Biochim. Biophys. Acta* **167**, 355 (1968).
43. H. Kimura, M. Mukai, and T. Kitamura, *J. Biochem.* (*Tokyo*) **76**, 1287 (1974).

44. A. Vandermeers, M. C. Vandermeers-Piret, J. Rathé, and J. Christophe, *Biochim. Biophys. Acta* **370**, 257 (1974).
45. B. Borgström and C. Erlandson, *Eur. J. Biochem.* **37**, 60 (1973).
46. A. Vandermeers, M. C. Vandermeers-Piret, J. Rathé, and J. Christophe, *FEBS Let.* **49**, 334 (1975).
47. B. Borgström, *J. Lipid Res.* **16**, 411 (1975).
48. O. Hernell and T. Olivecrona, *J. Lipid Res.* **15**, 367 (1974).
49. O. Hernell and T. Olivecrona, *Biochim. Biophys. Acta* **369**, 234 (1974).
50. O. Hernell, *Eur. J. Clin. Invest.* **5**, 267 (1975).
51. R. Tandon, R. H. Edmonds, and J. B. Rodgers, *Gastroenterology* **63**, 990 (1972).
52. W. S. Bradshaw and W. J. Rutter, *Biochemistry* **11**, 1517 (1972).
53. E. F. Curragh and D. T. Elmore, *Biochem. J.* **93**, 163 (1964).
54. A. M. Engel and B. Alexander, *Thromb. Diath. Haemorrh.* **27**, 594 (1972).
55. T. Exner and J. L. Koppel, *Biochim. Biophys. Acta* **321**, 303 (1973).
56. E. R. Cole, *Thromb. Diath. Haemorrh.* **32**, 132 (1974).
57. J. Barrowman and B. Borgström, *Gastroenterology* **55**, 601 (1968).
58. J. Barrowman, *Biochem. J.* **112**, 6P (1969).
59. M. T. Porter, A. L. Fluharty, S. D. de La Flor, and H. Kihara, *Biochim. Biophys. Acta* **258**, 769 (1972).
60. W. Mraz and H. Jatzkewitz, *Hoppe-Seyler's Z. Physiol. Chem.* **355**, 33 (1974).
61. A. L. Fluharty, R. L. Stevens, R. T. Miller, and H. Kihara, *Biochem. Biophys. Res. Commun.* **61**, 348 (1974).
62. R. L. Stevens, A. L. Fluharty, M. H. Skokut, and H. Kihara, *J. Biol. Chem.* **250**, 2495 (1975).
63. A. Jerfy and A. B. Roy, *Biochim. Biophys. Acta* **293**, 178 (1973).
64. A. L. Fluharty, R. L. Stevens, M. L. Scott, and H. Kihara, *Trans. Am. Soc. Neurochem.* **5**, 129 (1974).
65. D. A. Wenger, M. Sattler, C. Clark, and H. McKelvey, *Clin. Chim. Acta* **56**, 199 (1974).
66. H. Tanaka and K. Suzuki, *J. Biol. Chem.* **250**, 2324 (1975).
67. T. Miyatake and K. Suzuki, *J. Biol. Chem.* **250**, 585 (1975).
68. G. J. Walker and P. M. Hope, *Biochem. J.* **86**, 452 (1963).
69. M. D. O'Donnell, K. F. McGeeney, *Enzyme* **18**, 356 (1974).
70. M. D. O'Donnell, K. F. McGeeney, and O. Fitzgerald, *Enzyme* **19**, 129 (1975).
71. P. Emmelot, C. J. Bos, E. L. Benedetti, and P. H. Rumke, *Biochim. Biophys. Acta* **90**, 126 (1964).
72. P. Emmelot and C. J. Bos, *Biochim. Biophys. Acta* **99**, 578 (1965).
73. P. Emmelot and C. J. Bos, *Biochim. Biophys. Acta* **120**, 369 (1966).
74. P. Emmelot and C. J. Bos, *Biochim. Biophys. Acta* **150**, 341 (1968).
75. K. Konopka, M. Gross-Bellard, and W. Turski, *Enzyme* **13**, 269 (1972).
76. M. B. Lees, W. W. Sandler, and J. Eichberg, *Neurobiology* **4**, 407 (1974).
77. T. Parkinson and J. A. Olson, *Life Sci.* **3**, 107 (1964).
78. R. G. Faust and S. L. Wu, *J. Cell. Physio.* **67**, 149 (1966).
79. S. Hilden, H. M. Rhee, and L. E. Hokin, *J. Biol. Chem.* **249**, 7432 (1974).
80. A. F. Knowles and E. Racker, *J. Biol. Chem.* **250**, 3538 (1975).
81. B. K. Wallin and W. J. Arion, *Biochem. Biophys. Res. Commun.* **48**, 694 (1972).
82. H. M. Gunderson and R. C. Nordlie, *J. Biol. Chem.* **250**, 3552 (1975).
83. A. Ohkubo, N. Langerman, and M. M. Kaplan, *J. Biol. Chem.* **249**. 7174 (1974).
84. M. J. Frazer and J. H. Schulman, *J. Colloid Sci.* **11**, 451 (1956).
85. G. V. Vahouny and A. S. Brecher, *Arch. Biochem. Biophys.* **123**, 247 (1968).

86. B. Hadorn, J. Hess, B. Troesch, W. Verhaage, H. Gotze, and S. W. Bender, *Gastroenterology* **66**, 548 (1974).

87. L. Rutgeerts, G. Tytgat, E. Eggermont, *Tijdschr. Gastroenterol.* **15**, 379 (1972).

88. H. M. Kagam, G. D. Crombie, R. E. Jordon, W. Lewis, and C. Franzblau, *Biochemistry* **11**, 3412 (1972).

89. R. E. Jordan, N. Hewitt, W. Lewis. H. Kagan, and C. Franzblau, *Biochemistry* **13**, 3497 (1974).

90. L. H. Bernstein, S. Gutstein, and S. Weiner, *Proc. Soc. Exp. Biol. Med.* **132**, 1167 (1969).

91. R. K. Tompkins, J. F. Chow, and J. S. Clarke, *Am. J. Surg.* **124**, 207 (1972).

92. R. K. Tompkins, and R. M. Hayashi, *Am. J. Surg.* **128**, 633 (1974).

93. H. Theorell, S. Taniguchi, Ä. Äkeson and L. Skursky, *Biochem. Biophys. Res. Commun.* **24**, 603 (1966).

94. R. Pietruszko, *Biochem. Biophys. Res. Commun.* **60**, 687 (1974).

95. M. J. Lee and M. W. Whitehouse, *Biochim. Biophys. Acta* **100**, 317 (1965).

96. T. Urono, I. Takayanagi, K. Kubota, and K. Takagi, *Eur. J. Pharmacol.* **32**, 116, (1975).

97. R. F. Hanson, K. McCoy, and M. E. Dempsey, *Gastroenterology* **64**, 154 (1973) Abst.

98. L. Accatino and F. R. Simon, *Gastroenterology* **68**, 1067 (1975) Abst.

99. S. Makino, J. A. Reynolds, and C. Tanford, *J. Biol. Chem.* **248**, 4926, (1973).

100. J. B. Bawnik, R. Orda, and T. Wizniter, *Am. J. Dig. Dis.* **19**, 1143 (1974).

Chapter 4

BILE ACIDS IN HEPATO-BILIARY DISEASES

Helmut Greim

Abteilung für Toxikologie
Gesellschaft für Strahlen- und Umweltforschung
Neuherberg, Germany

I. INTRODUCTION

Bile acids are the end products of hepatic cholesterol metabolism. They are conjugated to taurine and glycine and are excreted with the bile into the intestine where they are reabsorbed and are again accumulated in the liver via the portal system. This enterohepatic circulation is well demonstrated but only little is known on the mechanisms involved in the transport of the bile acids through the membranes of the intestine and the liver cells during the uptake from the blood and excretion into the bile. Blood and bile are easily available and ample information exists on these two compartments of the bile acid pool, whereas less information is available on hepatic bile acids. Since almost all metabolic reactions of bile acid formation and conjugation are mediated by hepatic enzymes, changes in bile acid composition can be expected in liver diseases. Information on presently available data concerning such changes and the consequences thereof are presented subsequently.

II. BILE ACID METABOLISM IN HEPATIC DISEASES

Formation and metabolism of bile acids under physiological conditions have been reviewed by several authors (1–6). Those reactions of interest to the understanding of changes in bile acid composition and metabolism will be briefly recapitulated. The first step in the formation of bile acids is 7α-hydroxylation of cholesterol. Subsequently a 3-keto-derivative (7α-hydroxycholest-4-ene-3-one) is formed, which is a common precursor of the two primary bile acids, chenodeoxycholic acid and cholic acid. When the side chain of this intermediate is cleaved after 26-hydroxylation, a propionic acid-CoA intermediate is released and chenodeoxycholic acid results. Cholic acid is formed when 12α-hydroxylation precedes cleavage of the side chain, since only 7α-hydroxycholest-4-ene-3-one is a substrate for 12α-hydroxylation. After conjugation with glycine or taurine the bile acids are excreted into bile and passed into the gut, where they are subjected to deconjugation and C-7-dehydroxylation catalyzed by intestinal microorganisms. These reactions start in the distal section of the ileum and continue throughout the colon. Removal of the glycine or taurine conjugate is catalyzed by strains of clostridium, enterococcus, bacteroides, and lactobacillus; removal of the 7α-hydroxygroup by anaerobic, Gram-positive strains of lactobacillus results in the formation of the secondary bile acids. Dehydroxylation of cholic acid produces deoxycholic acid, dehydroxylation of chenodeoxycholic acid forms lithocholic acid. Most of the free or conjugated bile acids are reabsorbed. The main site of bile acid reabsorption is the distal part of the ileum by passive diffusion as well as by active processes. The pK of the taurine and the glycine conjugates is 1.5 and 3.5 respectively (7). This implies that at the pH of 6.2 in the gut the conjugates are almost completely in the ionic form. Generally, ionic hydrophilic compounds do not passively pentrate the lipophilic layer of cell membranes. Therefore, most of the intestinal reabsorption of bile acid conjugates should occur by active processes.

The pK of the deconjugated bile acids, however, is higher than the intestinal pH. Consequently, more than 50% of free bile acids are nonionized and are taken up by passive nonionic diffusion. During uptake, dihydroxylated bile acids have a greater affinity for the transport site than trihydroxylated ones. The same phenomenon has been found for the uptake of bile acids into the liver cells. This might be one explanation why trihydroxylated bile acid concentrations in the serum usually exceed those of the dihydroxylated acids. The monohydroxylated lithocholic acid is not soluble in water or in the lipid layer of the cell membranes. Although almost 300 mg of lithocholic acid are formed from chenodeoxycholic acid per day (8), only a minor proportion is reabsorbed, contributing only 3% of the bile acids in bile. Since many of the processes involved in the metabolism of bile acids are carried out by hepatic enzymes, it is evident that almost any pathological changes in liver function will affect bile acid metabolism.

A. Changes in the Steroid Nucleus

Until 15–20 years ago, general experience showed that there was an increase in serum bile acid levels in hepato-biliary disorders but there was no uniform consensus concerning changes in the pattern or quantity of individual bile acids. With improving analytical procedures it soon became evident that the ratio of the two main bile acids, chenodeoxycholic acid and cholic acid, occurs in approximately the same relative proportions in serum and in bile (9,10). In normal subjects, the ratio of cholic acid: chenodeoxycholic acid: deoxycholic acid is 1.0:0.92:0.88 (11,12,13). Since deoxycholic acid is formed from cholic acid in the gut, it can be expected that deoxycholic acid will decrease in concentration whenever cholestatic processes impair enterohepatic circulation of the bile acids. It has also become evident that bile acid determination in the serum in hepato-biliary disorders seems to separate the patients into two groups (14). In patients with predominantly hepatocellular injury, trihydroxy-: dihydroxy- ratio was less than one, whereas those with obstructive jaundice usually had a ratio greater than one. These observations have been confirmed by many investigators. Rudman and Kendall (15) found that the dihydroxy bile acids predominate in the sera of patients with Laennec cirrhosis, whereas patients with obstructive jaundice showed increased levels of cholic acid. Schiff (16), Carey (14), and Vlahcevic et al. (17) reported similar results.

More recent investigations, however, did not confirm these findings. Koruda and Okuda (11) showed increased cholic to chenodeoxycholic acid ratio in acute hepatitis and in patients with external bile fistula but a reduced ratio in chronic hepatitis, in cirrhosis, and in cholelithiasis. Paumgartner and Grabner (18) who investigated total amounts of conjugated bile acids found equal amounts of cholic acid and chenodeoxycholic acid in the serum of patients with intra- and extrahepatic cholestasis. In acute and chronic hepatitis and in liver cirrhosis chenodeoxycholic acid predominated. These data confirmed those obtained by Frosch (19) and Frosch and Wagner (20,21).

These contradictory results suggest that it may not be the *kind* of liver disease that affects cholic acid: chenodeoxycholic acid ratio but the severity of parenchymal dysfunction. This has already been suggested by Carey (14,22) and by Osborne et al. (23). Both groups of authors suggested that the occurrence of a reduced ratio of trihydroxylated: dihydroxylated (mostly chenodeoxycholic acid) bile acids concurrent with liver disease was a prognostically bad sign. This was confirmed by our own data (24). Studying hepatic bile acid concentration in patients with extrahepatic biliary obstruction, we found chenodeoxycholic acid to be the major bile acid only in prolonged cholestasis with bilirubin levels exceeding 20 mg %. Recent reports from the groups of Paumgartner (18) as well as of Korman (25) support this concept. Both groups of authors found any increases in bile acid levels, especially those in chenodeoxycholic and cholic acids in the serum, to be a very sensitive indicator for diagnosis of the beginning, as well as for the persistence, of a liver disease. Bile acid levels were found to be elevated in the serum before bilirubin increased in concentration, or they persisted longer than bilirubin elevation during recovery from liver dysfunction. Moreover, recurrence of a liver disease could be detected by control of bile acid levels before bilirubin or the transaminases became pathological.

The predominant rise in chenodeoxycholic acid concentration found in some hepatic diseases may be due to two mechanisms. One may be the specific inhibition of 12α-hydroxylase activity (6,26,27). This enzyme is situated in the microsomal fraction of the liver cell and determines formation of cholic acid and chenodeoxycholic acid from 7α-hydroxycholest-4-ene-3-one. If this precursor is 12α-hydroxylated, cholic acid is formed. If the side chain is cleaved before 12α-hydroxylation, chenodeoxycholic acid is formed. Thus, any inhibition of this reaction will increase chenodeoxycholic acid concentration. An additional mechanism possibly resulting in elevated levels of chenodeoxycholic acid during hepatic disorders may be increased 26-hydroxylation of cholesterol (28). This again would impair 12α-hydroxylation because of the introduction of the hydroxy group into the cholesterol side chain (6). Qualitative aspects of this pathway have to be determined. It seems to occur preferentially in the premature neonate (29) but it might be reactivated when bile acid synthesis via 7-hydroxycholesterol is impaired in liver diseases. Identification of 3β-hydroxy-5-cholenoic acid, another potential end product of cholesterol 26-hydroxylation, in the urine of patients with complete extrahepatic occlusion and with acute hepatitis (8) corroborates this assumption. The consequence of impaired 12α-hydroxylation or increased 26-hydroxylation seems to be reduced formation of cholic acid. Vlahcevic et al. (30,31) using labeled cholate in cirrhotics, demonstrated the cholate production to be reduced to one-third, whereas the total pool of bile acids was not affected in such patients. This again points to qualitative but not quantitative changes in the formation of bile acids from cholesterol at least in cirrhosis. Reduced formation of cholic acid contributes to a reduced pool of deoxycholic acid that is formed from cholic acid in the gut.

B. Changes in the Ratio of Glycine and Taurine Conjugates

Bile acids are excreted into bile as conjugates of taurine or glycine. In man this ratio of glycine:taurine conjugates in bile as well as in liver is about 3:1 (11,13). In liver diseases this ratio is altered because of the decrease in glycine conjugates. This was found in acute hepatitis, in cirrhosis, and in chronic aggressive hepatitis, as well as in biliary obstruction (11,13,18,20,21). Attempts to correlate the extent of these changes with the different liver diseases failed (18,20,21). Moreover, even among the patients with similar liver diseases the ratio varied largely so that changes in glycine:taurine ratio of bile acids cannot be used as a diagnostic criteria for liver diseases. Changes in the content of taurine in the food (12,32) may have more significance on glycine:taurine ratio than liver diseases.

Generally, liver diseases seem to affect glycine conjugation to a larger extent than conjugation of bile acids with taurine, which appears to be the preferential substrate of bile acid conjugation. Using rat liver homogenates, Bergström and Gloor (33) demonstrated that more than 90% of bile acids was conjugated in the presence of excess taurine while glycine had little effect. In human liver homogenates obtained from patients with cirrhosis, a relative reduction in glycine conjugation rate was seen (34). The immature liver of the newborn produces only taurine conjugates (35). One may assume, therefore, that glycine conjugation has more complex cofactor requirements, or may be less efficient, than conjugation with taurine. This may be why glycine conjugation is more strongly affected than taurine conjugation whenever liver function is impaired.

The clinical significance of the observed decrease of glycine:taurine ratio in liver disease is not clear. The main site of intestinal bile acid absorption is the distal part of the ileum, where conjugated and nonconjugated bile acids are reabsorbed against a concentration gradient (36–41). Small quantities of bile salts with higher lipid solubility, such as the glycine conjugates of the dihydroxy bile acids, are absorbed by passive nonionic diffusion (42) along the length of the intestine (43,44). Taurine conjugates, as well as glycine conjugates, have relatively low pK values of 1.5 and 3.5 respectively (7). At pH 6.4 in the ileum both conjugates are mostly nonionized and the passive nonionic diffusion rate should be similar for the two conjugates. Consequently, no change in reabsorption of bile acids should occur when the ratio of glycine:taurine conjugates is altered during liver disease. The major portion of bile acids, however, is absorbed by an active process in the distal part of the ileum (35,37,39–41, 45–48) where it rapidly passes through the mucosal cells to enter the portal system (49–54). There is evidence indicating the existence of a common transport site for which dihydroxy bile acids have a greater affinity for the transport site than trihydroxylated bile acids (55). It was observed in this study that taurocholic and cholic acid uptake was inhibited competitively in the presence of conjugated and nonconjugated bile acids. Similarly, Lack and Weiner (46), using everted sacs of the ileum, reported decreased absorption of a given bile acid in the presence of other bile acids. It was also found that trihydroxylated bile acids were absorbed better than dihydroxylated ones. Similar results have been obtained measuring the uptake of bile acids into liver cells, possibly explaining why dihydroxylated bile acid levels in the serum exceed levels of trihydroxylated ones (56). However, studies to compare absorption of glycine and taurine bile acid conjugates have not been done so far and the consequences of changes in glycine:taurine ratio occurring in hepatic diseases cannot be estimated.

C. Sulfation and Glucuronidation

Conjugation with the amino acids, taurine or glycine, increases water solubility of the bile acids and enables secretion into the bile (57). More recently, conjugation of the taurine and glycine derivatives with sulfate (58–64) and glucuronic acid (65) has been demonstrated. Conjugation with sulfate and glucuronic acid further increases water solubility of the original taurine and glycine conjugates. Sulfate esters were identified in the human bile, (58) in serum, and in the urine (59–64). Sulfate esters of lithocholic acid contribute 50–80% of total lithocholic acid found in bile of normal patients (58). In the urine of normal man only minute amounts of sulfated and nonsulfated bile acids were detected, whereas in a patient with intrahepatic bile duct hypoplasia, large quantities of bile salt sulfate esters of cholate and chenodeoxycholate were excreted in the urine (60). Moreover, in 40 patients with severe cholestasis due to extrahepatic obstruction, hepatitis, cirrhosis, and metastases of the liver—mono-, di-, and trisulfates of bile salts were identified in the urine—contributing to more than 50% of total bile salts excreted. By contrast, less than 10% of serum bile acids were found to be sulfated (64). Consequently, renal clearance of bile salt conjugates is higher than the clearance of the nonsulfated derivatives. Similar data were obtained by Back (8) who investigated patients with extrahepatic obstruction and with hepatitis. One can conclude from these observations that sulfation is a mechanism for elimination of bile salts during cholestatic liver injury. Similarly, glucuronidation of bile salts seem to be an escape mechanism facilitating excretion of bile acids during cholestatic diseases. In patients with intrahepatic and extrahepatic cholestasis approximately 20% and 10%, respectively, of the bile salts excreted in the urine were glucuronic acid conjugates (65). However, there were considerable individual differences, and in some patients the proportion of the glucuronides exceeded that of the sulfate esters.

It has been demonstrated that several UDP-glucuronyl transferases can be induced by phenobarbital (66,67). Since phenobarbital is known to reduce serum bile acid levels in patients with intrahepatic biliary atresia and in primary biliary cirrhosis by increasing fecal loss of bile acids (68,69), increased biliary excretion of bile acid glucuronides can be suggested.

As will be pointed out in Section III.A of this chapter, elevated serum concentration of monohydroxylated bile acids and dihydroxylated bile acids may be hepatotoxic. Hepatic sulfation and glucuronidation, forming more polar metabolites, enhance fecal and urinary excretion and may represent a protective mechanism in cholestasis.

D. Difference in Bile Acid Metabolism Between Laboratory Animals and Man

Many studies on bile acid metabolism have been performed in laboratory animals—rats and mice being most frequently used. However, bile acid metabolism in rats and mice differs from that in man. For example, in rat cholestasis, intrahepatic concentrations of cholate and β-muricholate greatly increase (70) and large amounts of β-muricholate are excreted in the urine (71). The trihydroxylated bile acids preferentially formed in rat cholestasis are water-soluble and less toxic than the monohydroxylated and dihydroxylated ones (72–76) and seem to be the rodent's escape mechanism for elimination of bile acids (70). Trihydroxylated β-muricholic acid is formed from chenodeoxycholic acid by 6β-hydroxylation via α-muricholic acid (77,78). This reaction is absent in man (79) who uses sulfation and glucuronidation to eliminate bile acids during cholestasis. The latter reactions are quantitatively less important in rodents (70). Although there is no doubt that experiments on bile acid metabolism performed in laboratory animals are necessary tools to understand pathways and regulation of bile acid metabolism, they do not necessarily represent the situation in man.

III. ROLE OF BILE ACIDS IN HEPATO-BILIARY DISORDERS

The physiological role of bile acids may be seen in the ability to form micelles in the bile to prevent precipitation of cholesterol, to emulsify lipid-soluble material in the gut, and to remove the products of lipolysis from the site of the enzyme to the microvillous membrane (80–82).

Relative changes in composition and concentration of biliary bile acids are involved in gallstone formation, which is discussed in Chapter 5 of this volume. Other consequences of altered bile acid patterns and concentrations in the gut, in the liver, and other organs are discussed subsequently.

A. Hepatocellular Necrosis

Hepatotoxic effects of lithocholic acid, deoxycholic acid, chenodeoxycholic acid, and cholic acid have been described. Ingestion of lithocholic acid induces duct cell proliferation and inflammatory changes in the livers of rats (83,84) and rabbits (85). Intravenous administration produces immediate cessation of bile flow (74,75,86) *in vivo* and in isolated rat livers *in vitro* (87). Since lithocholic acid is poorly water-soluble (88), the cholestatic effect may be attributed to precipitation of this bile acid which obstructs the biliary tract and results in elevation of serum bilirubin levels, and of activities of serum transaminases, and alkaline phosphatase (75). However, since lithocholic acid is almost insoluble in water, enteric reabsorption is minimal. Furthermore, in man as in the rat, lithocholic acid can be hydroxylated by microsomal enzymes (89) which further reduces the possibility of elevated levels in the liver. This may be the reason why increased concentrations could not be detected during intrahepatic cholestasis (90) or extrahepatic cholestasis (24) in human livers. Consequently, lithocholic acid does not seem to contribute significantly to hepato-biliary disorders (see also Section II.C of this chapter).

The dihydroxylated deoxycholic and chenodeoxycholic acids were shown to be hepatotoxic in *in vitro* experiments (72,91). At concentrations of 0.1 M found in rat livers after bile duct ligation, competitive inhibition of the microsomal enzyme system was demonstrated. This hepatic enzyme system metabolizes bile acids as well as foreign compounds such as drugs (92), explaining the old observations of prolonged hexobarbital narcosis of rats and rabbits in obstructive jaundice (93,94). It was further demonstrated by *in vitro* experiments that concentrations of chenodeoxycholic acid exceeding 0.3 mM are detergent (72,91). Microsomal cytochrome P-450, the key enzyme of the microsomal enzyme system (92), is converted to its inactive form, cytochrome P-420. Additionally, other microsomal membrane components are solubilized including cytochrome b-5 and NADPH cytochrome C reductase that are required for NADPH-mediated reduction of cytochrome P-450. Deoxycholic acid had similar effects in the *in vitro* experiments. However, intrahepatic concentration of this bile acid is not increased in liver diseases (90,95) and does not seem to contribute to the liver injury during hepatic disorders. By contrast, chenodeoxycholic acid is elevated in liver diseases (90,95). In mild cholestasis, sulfation and glucuronidation may prevent accumulation of this bile acid (58–65), but in severe cholestasis these escape mechanisms seem to be insufficient and chenodeoxycholic acid accumulates (24,90,95). Concentrations exceed 0.1 mM where competitive inhibition of drug metabolizing enzymes was demonstrated *in vitro* (72,91). In some patients intrahepatic chenodeoxycholic acid levels exceed 0.3 mM where detergent activity of this bile acid was observed *in vitro*. Correlation between high levels of dihydroxylated bile acids, mostly chenodeoxycholic acid, and liver cell necrosis–as demonstrated by feathery degeneration—suggest detergent capability of high chenodeoxycholic levels during cholestasis (24). These observations support the concept that increased hepatic bile acid levels are responsible for many of the functional and morphological changes in cholestatic livers (96).

In livers of patients with intrahepatic cholestasis the same pattern of bile acids have been found (90). Accordingly, similar consequences of increased bile acid concentrations can be expected in intra- and extrahepatic cholestasis, namely competitive inhibition of drug metabolizing enzymes and detergent effects.

The trihydroxylated cholic acid does not seem to be hepatotoxic in man. Although *in vitro* experiments show detergent effects at concentrations higher than 3 mM, and perfusion of rat livers with approximately 30 mM cholic acid solution induces ultrastructural changes in the hepatocytes (73), such high levels are not reached in liver diseases. Even in severe extrahepatic cholestasis, hepatic cholic acid concentration did not exceed 0.7 mM (24).

B. Inhibition of Drug Metabolism

In the preceding section it has been described that intrahepatic concentration of dihydroxylated bile acids, such as chenodeoxycholic and deoxycholic acids, are elevated in liver diseases and exceed levels of 0.1 mM. At this concentration microsomal enzymes that metabolize foreign compounds are competitively inhibited *in vitro*. In liver microsomes obtained from patients with acute hepatitis, Doshi *et al.* (97) showed reduced pentobarbital hydroxylase activity, and in severe liver diseases microsomal cytochome P-450 content was reduced (98). Consequently, the metabolism of drugs that are substrates of microsomal enzymes was expected to be inhibited during hepatic disorders.

The first attempts to substantiate this assumption gave negative results. Brodie *et al.* (99), investigating patients with chronic liver diseases, did not find changes in the elimination of phenylbutazone, aminopyrine, dicoumarol, and salicylic acid. Others observed elimination of diphenylhydantoin and phenobarbital (100) to be reduced in only a few of the patients investigated. However, in some patients with cirrhosis, metabolic inactivation of chloramphenicol was markedly decreased (101,102). Paraldehyde is metabolized mostly in the liver (103) and prolonged depression of the central nervous system was occasionally seen after standard doses in cir-rhotic patients (104). Ergot intoxication appears to be common in acute hepatic necrosis because of decreased detoxification (105), and tolbutamide produced prolonged hypoglycemic reactions in patients with cirrhosis (106,107), which can be attributed to a reduced elimination of the drug (108). Other hypoglycemic drugs such as glymidine behaved similarly in acute and chronic liver diseases (109,110). However, hypoglycemia is known to occur spontaneously in patients with hepatic disorders as well as in patients, without liver disease, who take hypoglycemic agents. In several patients with acute hepatitis, elimination of tolbutamide was increased (110) and was attributed to reduced binding of the drug to plasma proteins due to elevated bilirubin levels. Moreover, elimination of meprobamate (111), barbiturates, and phenazone (112–114) can be reduced.

Although present observations indicate that drug metabolism seems to be affected only in severe liver diseases, alteration in the disposition of drugs and undesirable drug effects should be generally expected in liver diseases. However, an abnormal drug response in hepatic disorders does not necessarily establish altered drug metabolism, since drug kinetics may be affected in other ways. Venous stasis and edema of the gut mucosa during portal hypertension may impair absorption (115) and the presence of ascites and edema can alter drug distribution (116). Moreover, other factors may add to undesired drug effects. Patients at the threshold of encephalopathia are very sensitive to sedatives (117), particularly morphine (118), and sensitivity to bis-hydroxycoumarol may be due to impaired synthesis and to binding of prothrombin to serum proteins (99,119).

C. Pruritus

In many hepatic disorders pruritus is one of the most disturbing symptoms. Pruritus is endured on all areas of the body, the palmares of the hands and feet being mostly affected (120). Pruritus associated with liver diseases has been thought to be related to increased serum bile acid concentrations (121). However, serum bile acid levels in patients with pruritus are not higher than in patients without itching (122,123). Moreover, when a patient suffering from alcoholic cirrhosis was fed cholic acid, serum cholic acid concentration rose to levels exceeding that of other patients with pruritus, but no itching occurred (14).

Schoenfield et al. (122) investigated the occurrence of bile acids in the skin of patients with pruritus and with high serum bile acid levels. Patients with pruritus had more bile acids in the skin than patients without pruritus.

This observation was recently confirmed by Stiehl (123), who demonstrated a close correlation between concentration of bile acids in the skin and pruritus.

Attempts to reduce pruritus have been made with cholestyramine (121,124,125). This ion-exchange resin is not absorbed after oral ingestion and binds bile acids, increasing fecal excretion, consequently lowering bile acid levels in the serum.

D. Malabsorption

Bile salts play an important role in the digestion and absorption of lipophilic food components. In a hydrophilic medium, lipids aggregate forming droplets of 4000–10,000 Å that present a small surface area for the attack of lipid splitting enzymes. In the gut, bile salts promote breakdown and absorption of the lipids by several different mechanisms (for review see 81,82,126). At first the bile salt conjugates excreted in the bile act as emulsifiers separating the large lipid aggregates into smaller fractions. This increases availability to lipase activity. Although some lipolytic activity is present in the stomach (127), the main lipolytic cleavage occurs in the small intestine by pancreatic lipase. Total secretory function of the pancreas, including secretion of bicarbonate and trypsin (128), as well as of lipase (129), is promoted by intraduodenal bile salts. Moreover, there is indication that bile salts also stimulate enteral gastrine release from the pyloric antrum (130,131), thus promoting gastric secretion.

In the gut, bile salts increase the activity of pancreatic lipase by two mechanisms: they shift the pH maximum from alkaline to weakly acid, and together with calcium ions, they increase water solubility of the liberated free fatty acids preventing a possible enzyme inhibition by the reaction products (126). Reabsorption of the insoluble products of lipolysis, such as long-chain fatty acids and monoglycerides is promoted by the detergent effect of the bile acids. Conjugated bile acids are detergents in that one part of the molecule is polar and water-soluble while the other part is lipophilic (132). When present above a critical micellar concentration the bile acids form polymolecular complexes called micelles. They are of spherical shape, having a diameter of 40–80 Å. The hydrophilic groups of the bile salts are oriented towards the external water face while the hydrophobic parts are directed to the inner space of the particle. Up to a certain concentration, further insoluble molecules such as long-chain fatty acids and monoglycerides, as well as cholesterol, can be incorporated (133) and thus be solublized.

After micellar solubilization the products of lipolysis diffuse primarily into the jejunal epithelium where they are reesterified and passed into the lymphatic system as chylomicrons (82,134). By contrast, medium-chain triglycerides do not require incorporation into micelles for reabsorption and pass readily into the portal vein (135). Although these lipids are negligible in the average American and European diet, they are useful for lipid substitution during bile acid deficient diseases with malabsorption syndromes (135–137).

Due to the central role of the bile acids in the digestion and reabsorption of lipids in the intestine, any bile acid deficiency or enteral loss of bile acids will reduce solubilization and reabsorption of the lipophilic food components. Consequently, some degree of steatorrhea very often is present in hepato-biliary disorders (138–140), resulting in malabsorption with impaired uptake of fat-soluble vitamins, calcium, or iron, which originates complex symptoms (141).

By far the most frequent cause of bile acid deficiency in the gut is a hepato-biliary disorder. Decreased formation of bile salts in acute viral hepatitis and in chronic liver diseases, such as cirrhosis, reduced reflux of absorbed bile salts due to portocaval shunt, and diminished excretion of bile salts into the bile in intrahepatic and extrahepatic cholestasis is presented in detail in Section IV of this chapter. Increased loss of bile salts in the blind loop syndrome and enhanced passage of the gut content after surgical jejuno-ileal shunting is discussed subsequently.

E. The Blind Loop Syndrome after Jejuno-Ileal Bypass

The role of bile acids in the pathogenesis of hepatic failure following jejuno-ileal bypass has not been fully clarified. This operation is well established in the treatment of severe obesity and leads to considerable weight loss without restriction of food intake (142–144). Originally the jejuno-cholic anastomosis was preferred (142, 145). For the last 20 years, however, the jejuno-ileal bypass has been selected because 5–10% of the patients with jejuno-cholic anastomosis developed chronic liver diseases (142,146). After jejuno-ileal bypass, chronic liver diseases occurred in 3.3% of 1400 cases described in the literature (147). Lethal liver failure was seen with 1.7% frequency. A common postsurgical consequence is the development of a fatty liver which exists in about 61% of the patients before surgery but in almost 100% after jejuno-ileal bypass (148). Generally, a malabsorption syndrome develops which resembles that after ileal resection with steatorrhea, diarrhea, and increased incidence of renal calculi and vitamin B_{12} deficiency (149). However, liver failure has not been seen after ileal resection, so that the liver disease developing after jejuno-ileal bypass seems to be due to the dysfunctionalized loop. One of the major features of the loop stasis developing after installation of the bypass seems to be bacterial overgrowth in the blind segment (149). This suggests that altered liver function and morphology after jejuno-ileal bypass may be due either to formation of bacterial toxins or to bacterial formation of hepatotoxic lithocholic acid in the segment. Although increased formation of lithocholic acid from chenodeoxycholic acid is expected to occur in the presence of bacterial overgrowth, no enhanced lithocholate levels were observed in the plasma of such patients, whereas plasma levels of chenodeoxycholate and of cholate were elevated in some of the patients with indices of hepatic failure (147,150). This again contradicts the hypothesis of increased reabsorption of lithocholate and supports the concept that bacterial toxins may be the culprit of the liver disease. This is further endorsed by observations in dogs with jejuno-ileal bypasses to which vibramycine was given. Whereas all animals in the group without the antibiotic showed increased fat infiltration of the liver and were dead within four months, all treated animals survived and no liver damage could be detected (151). This correlated with the absence of anaerobic bacteria in the blind loop of the treated dogs.

Another consequence of the bacterial overgrowth in the blind loop is an increased formation of free bile acids from taurine and glycine conjugates (152–156). Since free bile acids do not participate optimally in micelle formation (157), a decreased micellar solubilization of lipids occurs—resulting in fat malabsorption (158). Deconjugated bile acids have also been shown to impair small intestinal monosaccharide transport mechanism in experimental animals *in vitro* (159,160) and *in vivo* (161). This explains why secondary monosaccharide intolerance develops in patients with protracted diarrhea with increased bacterial bile salt deconjugation in the upper small intestine (162) and may add to the malabsorption syndrome after jejuno-ileal bypass. Furthermore, increased bacterial dehydroxylation of cholate (forming deoxycholate) was also suggested to contribute to a decrease in fat absorption (149). This would increase deoxycholate concentration in the gut, which *in vitro* inhibits reesterification of triglycerides in the mucosa (163). Whether bile salts significantly affect mucosal function *in vivo* has not been clarified (164–167). However, deoxycholate was found to be elevated in the plasma of some of the patients with blind loops after jejuno-ileal anastomosis (150) so that impaired reesterification may also contribute to the malabsorption of triglycerides in the disease.

F. Gallstone Formation

The factors involved in the formation of gallstones and therapeutic approaches to this disease will be described in Chapter 5 of this volume. In

this context the role of bile acids in the formation and dissolution of cholesterol gallstones that are most frequently seen in man are briefly discussed.
Solubility of cholesterol in bile largely depends on the biliary concentration
of bile salts and phospholipids forming mixed micelles with cholesterol. Any
changes in the biliary secretion of phospholipids or bile salts that increase
relative cholesterol concentration in the bile or any increased cholesterol
biosynthesis will result in supersaturation of cholesterol with consequent
precipitation and gallstone formation (168).

Formation of cholesterol is dependent on the activity of the
microsomal enzyme, 3-hydroxy-3-methylglutaryl CoA reductase (HMG
CoA reductase), forming mevalonic acid (169). Feedback control—by
reabsorbed cholesterol and by the enterohepatic circulation of bile salts—is
suggested as the means by which the activity of this enzyme is regulated
(170–174). Although a small fraction is metabolized to steroids in different
organs, most of the cholesterol is transformed to bile acids (173), which
occurs almost exclusively in the liver. The rate-controlling step is the 7α-
hydroxylation of cholesterol which in man is under feedback control of
cholesterol and bile acids (5). In patients with cholesterol gallstones the rate
of HMG CoA reductase was increased by 27%, whereas the activity of 7α-
hydroxylase of cholesterol was increased by 50% with a concomitant
increase in hepatic cholesterol content (174). The authors conclude from
this observation that the pathogenesis of gallstones is related to both
increased cholesterol synthesis and a reduced conversion of cholesterol into
bile acids. In gallstone carriers the bile acid pool size is reduced (175)
whereas hepatic cholesterol synthesis is found to be increased (176,177).
Consequently, formation of cholesterol stones may result from the inability
of the patients to increase bile salt synthesis in order to maintain sufficient
amounts in bile for micellar solubilization of cholesterol. Although the
effects of small bile acid pools may be overcome by increasing the frequency of intrahepatic circulation in gallstone patients (178), it is likely that
the hepatic bile salt concentration cannot be maintained. Moreover, a
reduced feedback inhibition of cholesterol 7α-hydroxylase by reduced intrahepatic bile salt levels will result in overproduction of cholesterol and
supersaturated bile. This may explain why in diseases with decreased bile
acid pool sizes such as cirrhosis or ileal dysfunction with reduced bile salt
reabsorption, the bile becomes lithogenic (26,30,179,180).

Feeding of chenodeoxycholic acid to patients inhibits hepatic HMG-
CoA-reductase and cholesterol production, thus rationalizing chenodeoxycholate therapy to dissolve cholesterol stones (181–183). Other
attempts to dissolve cholesterol stones were to increase relative phospholipid content in bile by feeding glycerophosphate or by phenobarbital
administration. Oral administration of glycerophosphate to rats increased

the lipid concentration in bile (184,185). When given to patients, solubility of cholesterol in bile was enhanced by relatively increased contents of phospholipids and bile acids (185). Phenobarbital in the monkey alters hepatic metabolism in a way that the bile produced is less saturated with cholesterol (186). In the hamster, however, phenobarbital induces gallstone formation (187).

IV. BILE ACID KINETICS, PATTERN, AND CONCENTRATION IN HEPATIC DISORDERS

Since bile acid synthesis from cholesterol and excretion into the gut is one of the major functions of the liver, any parenchymal disorders of the liver will affect the kinetics, pattern, and concentrations of the bile acids in blood, in bile, in the gut, and in the liver. Plasma bile acids have been measured mainly to obtain an index for differential diagnosis of the hepatic diseases. Increase of total plasma bile acid content has been reported by many investigators (9,14,15,23,85,188–193). With improving techniques, information on conjugated and nonconjugated bile salts as well as the detailed patterns of individual free and conjugated bile salts was obtained (193–196). Many authors suggested a close correlation between serum bile acid levels and serum bilirubin, which, however, is found only in severe hepatic dysfunction. Similarly, relation to other indices of liver diseases, such as serum alkaline phosphatase, 5′-nucleotidase, isocitric dehydrogenase, cholesterol, and serum glutamic oxalacetic transaminase (SGOT) exists only in severe hepatic dysfunction (18,193,197). In subclinical hepatic disorders plasma bile acids can be increased in concentration without bilirubin elevation (18). Moreover, bile acids increase before other diagnostic parameters for liver function change and were found to be very sensitive indicators for the persistence and recurrence of chronic active liver diseases (25; see also Section III. A of this chapter).

Whenever cholestasis results in impaired biliary secretion of bile acids, urinary excretion tries to compensate and bile acids appear in the urine. The mechanism of the renal excretion remains unclear. Investigations in dog (198) showed that cholate, as well as its glycine and taurine conjugates, undergo glomerular filtration and active reabsorption in the proximal tubule. In the distal tubule reabsorption was minor. Accordingly, Alström and Norman (199) suggested that renal excretion in the dog and possibly in man may be limited by two factors: (1) limited glomerular filtration due to extensive protein binding and (2) active reabsorption by the tubular cells. Consequently, increased amounts of serum bile acids would exceed binding capacity and would increase glomerular filtration. If, additionally, tubular reabsorption is blocked, which may occur in the presence of bile salt sulfates (200), increased amounts of bile acids could be excreted. Experimental results, however, do not fully support this concept. Even in patients with massive hepatic necrosis, urinary bile salt excretion did not exceed 5% ^{14}C-labeled cholic acid during the four days after administration (199). This is in contrast to the situation in infants with intrahepatic cholestasis where almost all injected labeled cholic acid appeared in the urine (201). Moreover, in 40 patients with extrahepatic obstruction, hepatitis, cirrhosis and metastases of the liver, total urinary excretion of lithocholate, deoxycholate, chenodeoxycholate, and cholate did not exceed 45 mg/24 h (64). Thus, in the light of increased bile salt levels in the blood of patients suffering from liver diseases, renal excretion does not seem to compensate fully for disturbed biliary elimination. More than 50% of all bile salts in the urine are sulfate esters (64). This agrees with the data of Back (8) who found even higher proportions of sulfated bile acids in the urine during acute hepatitis and biliary obstruction. One may conclude from this that sulfation is a precondition for renal elimination. However, most of the bile salt sulfate esters are found in bile (58,202). Consequently, enhanced sulfation and possibly glucuronidation observed in hepatic disorders (58–62,64,65,203–205) do not seem to protect from bile salt overload exclusively by enhanced renal excretion, but by excretion of the conjugates into the intestine. Because of a lower lipid solubility, the bile salt sulfates are not reabsorbed during intestinal passage and are eliminated in the feces (206). Desulfation of lithocholate occurs in healthy subjects in the distal intestine followed by reabsorption of the unsulfated molecule (205). However, there is complete fecal excretion of labeled lithocholic acid after several enterohepatic circulations because lithocholic acid becomes increasingly sulfated (205). This indicates that, at least in healthy subjects, hepatic sulfation exceeds bacterial desulfation resulting in a complete disposal of this bile salt in the feces. The subcellular site and the enzymes responsible for bile acid sulfation have not been identified. Since large amounts of sulfate esters are excreted during liver diseases, this enzymatic reaction seems to be maintained even in severe hepatic dysfunction.

Additionally, up to 50% of the bile salts excreted in the urine during acute hepatitis were shown to be water-soluble bile salt glucuronides (65). This contribution to renal bile salt elimination is even less understood.

The hepatic enzyme, UDP-glucuronyl-transferase, is involved in this reaction (65) and can be induced by phenobarbital (66,67,207). Since phenobarbital lowers serum bile salt concentration in patients with intrahepatic biliary atresia and primary biliary cirrhosis by increasing fecal excretion of bile salts (68,69), increased formation of bile salt glucuronides may be involved in this effect. Whether phenobarbital also induces sulfation of bile salts has not been investigated. In the rat, phenobarbital increases the liver weight and bile flow but not excretion of bile salts (168,208–210). This increase in bile salt independent bile formation exceeds the increase in liver weight and is suggested to be mediated by an activation of sodium transport processes by the canalicular membrane (211,212). Recent results on an increased content of Na,K-ATPase of plasma membranes rich in canaliculi after phenobarbital pretreatment supports this concept (213). These observations, however, do not explain why phenobarbital in cholestasis reduces serum bile acids by increasing fecal loss. Cholestasis may partially be overcome by stimulation of the bile salt independent bile flow, which at least in the rat, indirectly leads to increased secretion of bile salts (210).

A. Bile Acids for Diagnosis of Hepatic Disorders

Besides estimating total plasma bile acids, changes in the ratio of trihydroxy- to dihydroxy-acids in plasma were used to obtain information on the kind and severity of liver diseases. During the investigations of Carey it became evident (14) that patients with hepato-cellular injury and a trihydroxy- to dihydroxy acid ratio (cholic acid versus the sum of deoxycholic and chenodeoxycholic acids) of less than one, whereas in obstructive jaundice the ratio was greater than one. This has been confirmed by studies of Rudman and Kendall (15) and by Schiff (16). Persistence of a ratio less than one was frequently followed by coma, death, or both, whereas improvement of an acute or transient liver injury was paralleled by a rise in the ratio towards a value of one or more. This was regarded to be helpful in estimating the extent of liver injury. A similar correlation between the elevation in chenodeoxycholate relative to cholate concentration and the extent of liver cell injury was observed when concentration and distribution of intrahepatic bile acids were investigated (70,95). Since deoxycholate is frequently seen to disappear in hepatic disorders, its relative decrease was also investigated for diagnostic purposes, however, with limited success.

All these investigations have been performed using relatively difficult procedures (including extraction, derivatization, spectrophotometry, and gas chromatography) to determine bile salt concentrations in serum, bile, and liver. This complexity of the chemical reactions as well as difficulties in the quantitation did not permit use of these procedures in routine laboratories. A recent approach using a simple and sensitive radioimmunoassay to measure serum concentrations of conjugates of cholic acid seems to solve this problem (214). Using this technique, plasma conjugate levels (214) of cholic acid were compared with conventional biochemical tests in patients with chronic active liver diseases, based on histological diagnosis during treatment and during follow-up studies (25). The authors found changes in cholate levels to be more sensitive than conventional parameters such as total bilirubin and serum glutamic oxalacetic transaminase. At the beginning of treatment, cholate levels as well as SGOT activities, albumin, globulin, and sulfobromophthalein levels, and alkaline phosphatase activity, were increased. During the remission processes, which were achieved by appropriate doses of steroids with or without azathioprine, serum cholate levels were elevated longer than the conventional parameters. Moreover, when histological features of active hepatitis were no longer evident, serum cholate levels distinguished the patients into two groups: one with normal cholate levels without relapse within six months after remission and discontinuation of treatment, and another group with cholate levels three times normal after the six-month period. When relapse occurred during discontinuation of the treatment, cholate levels became abnormal before the SGOT activities changed, predicting the chemical and histological evidence of the active process. These determinations were done in fasting subjects. Another report suggests that determination of postprandial plasma bile acid levels is even more indicative of impaired liver function (215,216). Very recently, the radioimmunoassay technique was used to determine hepatic clearance of intravenously injected cholate (217), since delayed plasma disappearance of cholate has been shown in patients with hepatic or biliary tract diseases (218,219). The test proved to be very sensitive when applied to patients with chronic active or persistent liver diseases (220). In 36 patients, plasma disappearance of intravenously injected cholylglycine was compared to fasting levels of conjugated cholate and other conventional liver tests. In 25 patients with abnormal results of the conventional tests, 20 showed increased cholate levels, whereas plasma disappearance of cholate was delayed in all. Moreover, of the 11 patients with biochemical but not histological remission of the chronic active process, cholate levels were increased in 3 of 10, whereas cholate disappearance was delayed in 9 of the 11 patients.

B. Acute Hepatitis

For obvious reasons, only limited information exists on the mechanism of bile acid changes in acute hepatitis. In the acute phase total plasma bile acids are high, with concentrations more than a hundred times normal (193). At the same time the ratio of cholate:chenodeoxycholate:deoxycholate in bile, which is approximately 1:1:1 in control subjects (11–13) changes to 1:0.66:0.07. This marked reduction in deoxycholate was related to a reduced hepatic production of cholate (26). Diminished amounts of cholate secreted into bile would lead to almost complete reabsorption in the duodenum leaving no substrate for bacterial dehydroxylation to deoxycholate in the colon. Other factors such as reduced bacterial dehydroxylation of cholate, impaired absorption of deoxycholate, or reduced excretion of deoxycholate by the liver do not seem to contribute to reduced deoxycholate levels in bile. There is no proof that the bacterial population in the colon is changed during acute hepatitis nor is there increased deoxycholate excretion in the feces (221). A reduced capability of the liver to excrete deoxycholate would imply relatively high plasma levels, which have not been observed in acute hepatitis (14,19,20). However, cholic acid was found to disappear more slowly than normal from the serum of jaundiced patients with infectious hepatitis (222). This indicates impaired secretion of this trihydroxylated bile acid into the gut, which reduces bacterial production of deoxycholate during hepatitis. Moreover, cholic acid is excreted in the urine in acute hepatitis (199,222). This pathway seems to be used only in cholestasis since patients with biliary drainage and exhibiting pathological liver tests do not eliminate bile acids in the urine (199). Another possible mechanism could be increased transformation of deoxycholate to cholate. This reaction occurs in the rat (223), but man is not able to hydroxylate deoxycholate at the 7-α position (224).

C. Chronic Hepatitis and Cirrhosis

Although ample information exists on changes in serum bile acid levels during acute hepatitis, most of the work on the kinetics and pattern of bile acids has been done in chronic hepatitis and cirrhosis. Besides the observation of increased plasma bile acid levels and a very low percentage of biliary deoxycholate (17) which also occurs in acute hepatitis, it was the diminished intrahepatic bile acid pool, the decreased cholic acid synthesis, the reduced deoxycholate levels in plasma and bile, and a relative increase in plasma dihydroxylated bile salts which drew much attention in patients with chronic liver diseases. Plasma bile acid levels seem to be elevated in most of the cirrhotic patients (14,15,23,189,218) although cholestasis does not seem to be a major consequence of cirrhosis. There is evidence, however, that substantial amounts of bile acids undergo portal systemic shunting after reabsorption and enter systemic circulation without passing through the liver. This was suggested from observations that healthy individuals to whom [^{14}C]cholate had been injected did not respond with a rise in serum radioactivity after one intrahepatic circulation, whereas cirrhotic patients responded with a spontaneous rise in radioactivity (225).

The observation of a diminished intrahepatic bile acid pool in cirrhotics has been made during attempts to explain the diminished contents of deoxycholate and lithocholate in bile (17). The markedly decreased bile acid pool in cirrhotic patients (26,30), which can be reduced by as much as 50% (179), seems to be due mostly to reduced cholate synthesis. In patients without liver diseases cholic acid synthesis was about twice that of chenodeoxycholate. In patients with cirrhosis this difference was abolished. Since chenodeoxycholate synthesis is not affected, or affected only to a lesser extent, the site of effect may be 12α-hydroxylation of 7α-hydroxycholest-4-ene-3-one. (See Section II.A of this chapter). A consequence of the block of cholic acid synthesis is the lack of deoxycholic acid in bile (13,17) and plasma (179,226), which may be further aggravated by a decreased enterohepatic circulation of cholic acid due to cholestasis. Recent observations in cirrhotic patients support this concept of multifactorial decrease of deoxycholate levels (227). When sufficient amounts of cholate were fed to ten cirrhotic patients, seven responded with an elevation of biliary deoxycholate, which returned to normal in three of them. However, in three patients with negligible biliary deoxycholate levels, the feeding of cholate had little or no effect, although the amount of cholate given on each of four consecutive days corresponded to the normal total pool of this bile acid. Consequently, a deficiency of cholic acid in the intestine could not be the reason for reduced biliary deoxycholate levels.

Reduced synthesis of cholate relative to chenodeoxycholate and the disappearance of deoxycholate in bile and serum during cirrhosis were used as an index of the severity of the liver disease. When McCormick et al. (26) investigated the synthesis of chenodeoxycholic acid and cholic acid in cirrhotic patients, they observed no effect on chenodeoxycholate synthesis but a reduced cholate formation. In addition, there was a high degree of correlation between cholate synthesis and the severity of the disease, which was assessed by clinical and by laboratory parameters including liver biopsy. Cholate synthesis was markedly reduced in the group of patients with mild cirrhosis. Although the synthesis rate was further reduced as the disease advanced, the greatest decrease was seen during the earlier phases of the disease. The authors conclude from this observation that significant alterations in cholate synthesis must occur before the onset of clinical symptoms.

D. Intra- and Extrahepatic Cholestasis

Cholestasis is a bile flow stagnation due to a failure of biliary secretion of the liver cells (96,228,229) with a concomitant accumulation in the blood of the constituents normally excreted in bile. From the etiological point of view one may differentiate between extrahepatic factors that block bile flow into the duodenum, such as common-duct stones, cancer of the biliary tract or of the pancreas, and factors that cause intrahepatic cholestasis. The latter form may be further differentiated pathogenetically into conditions where the primary lesion is located inside the liver cell such as cholestatic viral hepatitis, by altered membrane function due to steroids, by lithocholate or chlorpromazine precipitates in the canaliculus, by intralobular bile duct obstruction during primary biliary cirrhosis, or in sclerosing cholangitis and cholangiocarcinoma, when bile flow in segmental or larger intrahepatic bile ducts is impaired (229).

In extrahepatic cholestasis bile salts highly increase in concentration in the liver and in plasma (24). In the liver, cholic acid increases primarily during onset of the obstruction whereas in prolonged cholestasis with increased bilirubin levels, chenodeoxycholate becomes highly elevated too. As has been described in Section III.A of this chapter, high dihydroxy bile salt levels—mainly chenodeoxycholate levels—aggravate the disease by detergent activity. Hepatic content of deoxycholate is largely reduced due to the interrupted intrahepatic circulation. Intrahepatic levels of bile salt conjugates of taurine, glycine, or sulfate esters have not been determined.

Ample information exists on concentration and pattern of bile salts in the plasma of patients with cholestasis. Generally, depending on the severity of the disease, patients with extrahepatic biliary obstruction show the same plasma bile acid pattern as patients with intrahepatic cholestasis (90,230) and no information for differential diagnosis can be obtained. This also indicates that changes in bile acid pattern in cholestasis are a consequence, rather than a cause, of the obstruction. A close correlation, however, exists between the extent of cholestasis and the cholate:chenodeoxycholate ratio. Carey (85) suggested that a low ratio indicates a bad prognosis. The limit was set at 1.0, whereas van Berge Henegouwen and Brandt set it at 0.7 (231). Determination of bile acids in the liver of patients with obstructive jaundice did not fully confirm these suggestions (70). In short-termed cholestasis, cholic acid was predominantly elevated. In prolonged biliary obstruction with high serum bilirubin levels, chenodeoxycholic acid as well as cholic acid rose, but chenodeoxycholic acid levels never exceeded those of cholic acid. One has to consider, however, that estimation of bile acids in liver biopsies not only includes intracellular bile acids but also those in the hepatic blood and in intrahepatic bile. When necropsy livers were studied, less cholic acid but more chenodeoxycholic acid than in biopsy specimens with similar diseases were found. Determination of bile acids in necropsy livers, therefore, may not reveal a true pattern of bile acid distribution during life.

Quantitative aspects of the changes in bile acid metabolism during cholestasis are determined by reduced synthesis and by a bile acid pool restraint to the liver. In rhesus monkeys and man, complete obstruction leads to a total arrest of primary bile acid synthesis in the liver (232,233). Since bile acid synthesis is controlled by feedback regulation of 7β-hydroxylation of cholesterol (5,234), any reduction in activity by high intrahepatic bile acid levels would help to protect the hepatocyte from toxic bile acid overload (26). Preferential formation of cholate in the beginning of obstruction as well as sulfation and glucuronidation further add to this self protection. Under these circumstances the bile acid pool is largely restrained to the liver with larger portions being distributed in the systemic circulation, the extracellular space, and the skin.

A very special form of biliary obstruction is the extrahepatic biliary atresia of infants. These children without bile ducts excrete all bile salts via the kidney. Large amounts of conjugated cholic and chenodeoxycholic acids of 3β-hydroxy-5-cholenoic acid and small amounts of lithocholic acid are found in the urine (233,235,336). Whereas cholic acid is mostly excreted as its glycine conjugate, chenodeoxycholic acid appears in the urine as a monosulfate of tauro- or glycochenodeoxycholate (237). Lithocholate is almost entirely in the form of its tauro-or glycolithosulfate ester. The 3β-hydroxy-5-cholenoic acid may also be sulfated (29,238). Since lithocholic acid, 3β-hydroxy-5-cholenoic acid, and chenodeoxycholic acid are found to be hepatotoxic (24,74,76,86), and the sulfate esters of tauro- and glyco-lithocholate have a weaker cholestatic effect than the original conjugates (76), sulfation serves as an important mechanism to protect from the toxic effects of these bile acids.

REFERENCES

1. H. Danielsson and K. Einarsson, *in* "The Biological Basis of Medicine" (E. E. Bittar and N. Bittar, eds.), p. 279, Academic Press, London (1969).
2. S. Linstedt, *in* Atherosclerosis: Proceedings of the Second International Symposium (R. J. Jones, ed.), p. 262, Springer-Verlag, New York (1970).
3. H. Greim, D. Trülzsch, P. Czygan, F. Hutterer, F. Schaffner, and H. Popper, *Ann. N.Y. Acad. Sci.* **212**, 139 (1973).
4. I. W. Percy-Robb and G. S. Boyd, *Scott. Med. J.* **18**, 166 (1973).
5. E. H. Mosbach and G. Salen, *Digestive Diseases* **19**, 920 (1974).
6. J. Bjorkhem and H. Danielsson, *Mol. Cell. Biochem.* **4**, 79 (1974).
7. D. M. Small, *in* "The Bile Acids: Chemistry, Physiology and Metabolism" (P. P. Nair and D. Kritchevsky, eds.), Vol. 1, p. 249, Plenum Press, New York (1971).
8. P. Back, *Z. Gastroenterol.* **11**, 477 (1973).
9. J. B. Carey, J. Figen, and C. J. Watson, *J. Lab. Clin. Med.* **46**, 802 (1955).
10. J. B. Carey, *Science* **123**, 892 (1956).

11. T. Koruda and K. Okuda, *Acta Hepato-Gastroenterol.* **21**, 120 (1974).
12. S. Lindstedt, D. S. Avigan, J. Sjövall, and D. Steinberg, *J. Clin. Invest.* **44**, 1754 (1965).
13. J. Sjövall, *Clin. Chim. Acta* **5**, 33 (1960).
14. J. B. Carey, *J. Clin. Invest.* **37**, 1494 (1958).
15. D. Rudman and F. E. Kendall, *J. Clin. Invest.* **36**, 530 (1957).
16. L. Schiff, in "Hepatitis Frontiers" (F. W. Hartman, G. A. Logrippo, J. G. Mateer, and J. Barron, eds.), p. 31, Little, Brown and Co., Boston (1957).
17. Z. R. Vlahcevic, J. Buhac, C. C. Bell, and L. Swell, *Gut* **11**, 420 (1970).
18. G. Paumgartner and G. Grabner, *Acta Hepato-Splenol.* **23**, 344 (1970).
19. B. Frosch, *Arzneim.-Forsch.* **15**, 178 (1965).
20. B. Frosch and H. Wagner, *Dtsch. Med. Wochenschr.* **93**, 1754 (1968).
21. B. Frosch and H. Wagner, *Klin. Wochenschr.* **46**, 913 (1968).
22. J. Carey, *Medicine* **45**, 461 (1966).
23. E. C. Osborne, J. D. P. Wootton, L. C. Da Silva, and S. Sherlock, *Lancet* **2**, 1049 (1959).
24. H. Greim, D. Trülzsch, P. Czygan, J. Rudick, F. Hutterer, F. Schaffner, and H. Popper, *Gastroenterology* **63**, 846 (1972).
25. M. G. Korman, A. F. Hofmann, and W. H. J. Summerskill, *N. Engl. J. Med.* **290**, 1399 (1974).
26. W. C. McCormick, C. C. Bell, L. Swell, and Z. R. Vlahcevic, *Gut* **14**, 895 (1973).
27. H. Danielsson, *Steroids* **22**, 567 (1973).
28. K. E. Anderson, E. Kok, and N. B. Javitt, *J. Clin. Invest.* **51**, 112 (1972).
29. P. Back and K. Ross, *Hoppe-Seyler's Z. Physiol. Chem.* **354**, 83 (1973).
30. Z. R. Vlahcevic, J. Buhac, J. T. Farrar, C. C. Bell, and L. Swell, *Gastroenterology* **60**, 491 (1971).
31. Z. R. Vlahcevic, J. R. Miller, J. T. Farrar, and L. Swell, *Gastroenterology* **61**, 85 (1971).
32. J. Sjövall, *Proc. Soc. Exp. Biol. Med.* **100**, 676 (1959).
33. S. Bergström and U. Gloor, *Acta Chem. Scand.* **8**, 1373 (1954).
34. H. Ekdahl, *Acta chir. Scand.* **115**, 208 (1958).
35. J. C. Encrantz and J. Sjövall, *Clin. Chim. Acta* **4**, 793 (1959).
36. R. D. Baker and G. W. Searle, *Proc. Soc. Exp. Biol. Med.* **105**, 521 (1960).
37. L. Lack and I. M. Weiner, *Am. J. Physiol.* **200**, 313 (1961).
38. I. M. Weiner and L. Lack, *Am. J. Physiol.* **202**, 155 (1962).
39. P. R. Holt, *Am. J. Physiol.* **207**, 1 (1964).
40. M. R. Playoust and K. J. Isselbacher, *J. Clin. Invest.* **43**, 467 (1964).
41. J. E. Glasser, I. M. Weiner, and L. Lack, *Am. J. Physiol.* **208**, 359 (1965).
42. J. M. Diamond, E. M. Wright, and R. Whittam, *Ann. Rev. Physiol.* **31**, 581 (1970).
43. J. M. Dietschy, H. S. Salomon, and M. D. Siperstein, *J. Clin. Invest.* **45**, 832 (1971).
44. J. M. Dietschy, *J. Lipid Res.* **9**, 297 (1968).
45. L. Lack and I. M. Weiner, *Fed. Proc.* **22**, 1334 (1964).
46. L. Lack and I. M. Weiner, *Am. J. Physiol.* **210**, 1142 (1966).
47. L. Lack and I. M. Weiner, *Gastroenterology* **52**, 282 (1967).
48. H. Buchwald and R. L. Gebhard, *Ann. Surg.* **167**, 191 (1968).
49. B. Josephson, *Physiol. Rev.* **21**, 463 (1941).
50. B. Josephson and A. Rydin, *Biochem. J.* **30**, 2224 (1936).
51. J. Sjövall and I. Akesson, *Acta Physiol. Scand.* **34**, 273 (1955).
52. T. Olivecrona and J. Sjövall, *Acta Physiol. Scand.* **46**, 284 (1959).
53. T. Chronholm and J. Sjövall, *Eur. J. Biochem.* **2**, 375 (1967).
54. R. T. Reinke and J. D. Wilson, *Clin. Res.* **15**, 242 (1967).
55. P. R. Holt, *Am. J. Physiol.* **210**, 635 (1966).

56. L. R. Schwarz, R. Burr, M. Schwenk, E. Pfaff, and H. Greim, *Eur. J. Biochem.* **55**, 617 (1975).
57. P. Ekwall, T. Rosendahl, and A. Sten, *Acta Chem. Scand.* **12**, 1622 (1958).
58. R. H. Palmer and M. G. Bolt, *J. Lipid Res.* **12**, 671 (1971).
59. A. Stiehl, M. M. Thaler, and W. H. Admirand, *Hormones Metabol. Res.,* Suppl. **4**, 49 (1974).
60. A. Stiehl, D. L. Earnest, and W. H. Admirand, *Gastroenterology* **68**, 534 (1975).
61. A. Stiehl, *in* "Bile Acids in Human Diseases" (P. Back and W. Gerok, eds.), p. 73, F. K. Schattauer, Stuttgart, New York (1972).
62. P. Back, J. Sjövall, and K. Sjövall, *Scand. J. Clin. Lab. Invest.* **29** (Suppl. 126), 17.12 (1972).
63. J. Makino, *Lipids* **8**, 47 (1973).
64. A. Stiehl, *Eur. J. Clin. Invest.* **4**, 59 (1974).
65. W. Fröhling and A. Stiehl, *Eur. J. Clin. Invest.* **6**, 67 (1976).
66. K. W. Bock, W. Fröhling, H. Remmer, and B. Rexer, *Biochim. Biophys. Acta* **327**, 46 (1973).
67. K. W. Bock and W. Fröhling, *Arch. Pharmacol.* **277**, 103 (1973).
68. A. Stiehl, W. H. Admirand, and M. M. Thaler, *N. Engl. J. Med.* **286**, 858 (1972).
69. A. Stiehl, M. M. Thaler, and W. H. Admirand, *Pediatrics* **51**, 992 (1973).
70. H. Greim, D. Trülzsch, J. Roboz, K. Dressler, P. Czygan, F. Hutterer, F. Schaffner, and H. Popper, *Gastroenterology* **63**, 837 (1972).
71. V. A. Ziboh, J. T. Matschiner, E. A. Doisy Jr., S. H. Hsia, W. H. Elliott, S. Thayer, and E. A. Doisy, *J. Biol. Chem.* **236**, 387 (1961).
72. F. Hutterer, H. Denk, P. G. Bacchin, J. B. Schenkman, F. Schaffner, and H. Popper, *Life Sci.* **9**, 877 (1970).
73. C. L. Witzleben, *Exp. Mol. Pathol.* **16**, 47 (1972).
74. N. B. Javitt, *Nature* **210**, 1263 (1966).
75. B. G. Priestly, M. G. Côté, and G. L. Plaa, *Can. J. Physiol. Pharmacol.* **49**, 1078 (1971).
76. M. M. Fisher, R. Magnusson, and K. Miyai, *Lab. Invest.* **21**, 88 (1971).
77. B. Samuelsson, *Acta Chem. Scand.* **13**, 976 (1959).
78. W. Voigt, P. J. Thomas, and S. L. Hsia, *J. Biol. Chem.* **243**, 3493 (1968).
79. J. Björkhem, K. Einarsson, and G. Hellers, *Eur. J. Clin. Invest.* **3**, 495 (1973).
80. J. R. Senior, *J. Lipid Res.* **5**, 495 (1964).
81. A. F. Hofmann, *in* "Handbook of Physiology" (C. C. Code, ed.), p. 2507, Williams and Wilkins, Baltimore (1968).
82. J. M. Johnston, *in* "Handbook of Physiology" (C. C. Code, ed.), p. 1353, Williams and Wilkins, Baltimore (1968).
83. R. D. Hunt, G. A. Léveillé, and H. E. Sauberlich, *Proc. Soc. Exp. Biol. Med.* **115**, 227 (1964).
84. K. Miyai, W. W. Mayr, and A. L. Richardson, *Lab. Invest.* **32**, 527 (1975).
85. J. B. Carey, *in* "Diseases of the Liver" (L. Schiff, ed.), p. 103, Lippincott and Co, Philadelphia (1969).
86. N. B. Javitt and S. Emerman, *J. Clin. Invest.* **47**, 1002 (1968).
87. K. Miyai, V. M. Price, and M. M. Fisher, *Lab. Invest.* **24**, 292 (1971).
88. D. M. Small and W. Admirand, *Nature* **221**, 265 (1969).
89. D. Trülzsch, J. Roboz, H. Greim, P. Czygan, J. Rudick, F. Hutterer, F. Schaffner, and H. Popper, *Biochem. Med.* **9**, 158 (1974).
90. H. Greim and P. Czygan, *Verh. Dtsch. Ges. Inn. Med.* **80**, 443 (1974).
91. F. Hutterer, P. Bacchin, H. Denk, J. B. Schenkman, F. Schaffner, and H. Popper, *Life Sci.* **9**, 1159 (1970).

92. F. Hutterer, H. Greim, D. Trülzsch, P. Czygan, and J. B. Schenkman, in "Progress in Liver Diseases" (H. Popper and F. Schaffner, eds.), Vol. 4, p. 151, Grune and Stratton, New York, London (1972).
93. E. F. McLeun and J. R. Fouts, J. Pharmacol. Exp. Ther. 131, 7 (1961).
94. B. A. Becker and G. L. Plaa, Toxicol. Appl. Pharmacol. 7, 680 (1965).
95. H. Greim, P. Czygan, F. Schaffner, and H. Popper, Biochem. Med. 8, 280 (1973).
96. H. Popper and F. Schaffner, Hum. Pathol. 1, 1 (1970).
97. J. Doshi, A. Luisada-Opper, and C. M. Leevy, Proc. Soc. Exp. Biol. Med. 140, 492 (1972).
98. B. Schoene, R. A. Fleischmann, H. Remmer, and H. F. v. Oldershausen, Eur. J. Clin. Pharmacol. 4, 65 (1972).
99. B. B. Brodie, B. B. Burns, and M. Weiner, Med. Exp. 1, 290 (1959).
100. H. W. Kutt, W. Winters, R. Shennan, and F. McDowel, Arch. Neurol. 11, 649 (1964).
101. C. M. Kunin, A. J. Glatzko, and M. Finland, J. Clin. Invest. 38, 1498 (1959).
102. L. G. Suhrland and A. S. Weisgerber, Arch. Intern. Med. 112, 747 (1963).
103. H. Levine, A. J. Gilbert, and M. Bodansky, J. Pharmacol. Exp. Ther. 69, 316 (1940).
104. S. Sherlock, in "Diseases of the Liver", Davis & Co., Philadelphia (1963).
105. M. J. Whelton, A. Allaway, A. Stewart, and L. Kreel, Gut 9, 287 (1968).
106. H. J. Cohn, M. Perlmutter, J. N. Silverstein, and M. Numeroff, J. Clin. Endocrinol. 24, 28 (1964).
107. R. J. Samson, C. Trey, A. H. Timme, and S. J. Saunders, Gastroenterology 53, 291 (1967).
108. U. Ueda, T. Sakurai, M. Ota, A. Nakajama, K. Kamii, and H. Maezawa, Diabetes 12, 414 (1963).
109. H. Held, B. Kaminsky, and H. F. v. Oldershausen, Diabetologica 6, 386 (1970).
110. H. Held, H. Eisert, and H. F. v. Oldershausen, Arzneim.-Forsch. 23, 1801 (1973).
111. H. Held and H. F. v. Oldershausen, Klin. Wochenschr. 47, 78 (1969).
112. F. H. Franken and W. Hagelskamp, Dtsch. Med. Wochenschr. 95, 1613 (1970).
113. D. D. Breimer, W. Zilly, B. Keller, and E. Richter, Clin. Pharmacol. Ther. 18, 433 (1975).
114. J. Elfström and S. Lindgren, Eur. J. Clin. Pharmacol. 7, 467 (1974).
115. I. F. Enquist, M. R. Golding, R. G. Aiello, S. M. Fierst, and N. A. Salomon, Surg. Gynecol. Obstet. 120, 87 (1965).
116. G. A. Marin, M. L. Clark, and J. R. Senior, Ann. Intern. Med. 69, 1155 (1968).
117. A. J. Levi, in "Therapeutic Agents and the Liver" (N. McIntyre and S. Sherlock, eds.), p. 105, Blackwell, Oxford (1965).
118. J. Laidlaw, A. E. Read, and S. Sherlock, Gastroenterology 40, 389 (1961).
119. E. H. Reisner, J. Norman, W. W. Field, and R. Brown, Am. J. Med. Sci. 217, 445 (1949).
120. J. H. Henderson, Arch. Intern. Med. 130, 632 (1972).
121. T. B. van Itallie, S. A. Hashim, R. S. Crampton, and D. M. Tennent, N. Engl. J. Med. 265, 469 (1961).
122. L. J. Schoenfield, J. Sjövall, and E. Permann, Nature 213, 93 (1967).
123. A. Stiehl, Z. Gastroenterol. 12, 121 (1974).
124. Z. H. Oster, E. A. Rachmilewitz, E. Moran, and Y. Stein, Israel J. Med. Sci. 1, 599 (1965).
125. F. Scnairner, F. M. Klion, and A. J. Latuff, Gastroenterology 48, 293 (1965).
126. P. Desnuelle, in "Handbook of Physiology" p. 403, (C. C. Code, ed.), Williams & Wilkins, Baltimore (1968).
127. M. Cohen, R. G. H. Morgan, and A. F. Hofmann, Gastroenterology 60, 1 (1971).

128. K. G. Wormsley, *Lancet,* p. 586 (1970).
129. M. M. Forell, H. Stahlheber, and F. Scholz, *Dtsch. Med. Wochenschr.* **90,** 1128 (1965).
130. B. S. Bedi, H. T. Debas, G. Gillespie, and J. E. Gillespie, *Gastroenterology* **60,** 256 (1971).
131. H. Brunner, T. C. Northfield, A. F. Hofmann, V. L. W. Go, and W. H. J. Summerskill, *Mayo Clin. Proc.* **49,** 851 (1974).
132. A. F. Hofmann, *Gastroenterology* **48,** 484 (1965).
133. M. C. Carey and D. M. Small, *Am. J. Med.* **49,** 590 (1970).
134. W. O. Dobbins, *Am J. Clin. Nutr.* **22,** 257 (1969).
135. N. J. Greenberger and T. G. Skillman, *N. Engl. J. Med.* **280,** 1045 (1969).
136. R. B. Zurier, R. G. Campbell, S. A. Hashim, and T. B. van Itallie, *N. Engl. J. Med.* **274,** 490 (1966).
137. N. J. Greenberger, R. D. Ruppert, and M. Tzagournis, *Ann. Intern. Med.* **66,** 727 (1967).
138. E. Baraona, H. Orrego, O. Fernandez, E. Amenabar, E. Maldonado, F. Tag, and A. Salinas, *Am. J. Dig. Dis.* **7,** 318 (1962).
139. W. G. Linscheer, J. F. Patterson, E. W. Moore, R. J. Clermont, R. C. Sander, S. J. Robins, and T. C. Chalmers, *J. Clin. Invest.* **45,** 1317 (1966).
140. M. S. Losowsky and B. E. Walker, *Gastroenterology* **56,** 589 (1969).
141. H. J. Wildgrube and W. Siede, *Acta Hepato-Gastroenterol.* **21,** 151 (1974).
142. J. H. Payne, L. T. de Wind, and R. R. Commons, *Am. J. Surgery* **106,** 273 (1963).
143. J. H. Payne, and L. T. de Wind, *Am. J. Surg.* **118,** 141 (1969).
144. B. Husemann, *Dtsch. Med. Wochenschr.* **98,** 2343 (1973).
145. A. L. Lewis, R. B. Turnbull. and J. H. Page, *J. Am. Med. Assoc* **182,** 187 (1962).
146. D. B. McGill, S. R. Humphreys, A. H. Baggenstoss, and R. Dickson, *Gastroenterology* **63,** 872 (1972).
147. W. Schmeisser, C. P. Schrader, H. Greim, and H. F. von Oldershausen, *A. Gastroenterol.* **13,** 619 (1975).
148. W. Salmon, *Surg. Gynecol. Obstet.* **132,** 965 (1971).
149. D. Fromm, *Surgery* **73,** 639 (1973).
150. H. P. Sherr, P. P. Nair, J. G. Banwell, J. J. White, and D. H. Lockwood, *Gastroenterology* **64,** A-117 (1973).
151. J. P. O'Leary, J. W. Maher, J. I. Hollenbeck, and E. R. Woodward, Digestive Disease Week, A-236, San Francisco (1974).
152. S. A. Broitman and R. A. Giannella, *in* "Absorption Phenomena: Topics in Medicinal Chemistry" (J. L. Rabinowitz and R. M. Myerson, eds.), John Wiley & Sons, New York (1971).
153. M. J. Hill and B. S. Drasar, *Gut* **9,** 22 (1968).
154. B. S. Drasar and M. Shiner, *Gut* **10,** 812 (1969).
155. S. L. Gorbach and S. Tabaqchali, *Gut* **10,** 963 (1969).
156. R. M. Donaldson, *Adv. Intern. Med.* **16,** 191 (1970).
157. A. F. Hofmann and B. Borgström, *Fed. Proc.* **21,** 43 (1962).
158. Y. S. Kim, N. Spritz, M. Blum, J. Terz, and P. Sherlock, *J. Clin. Invest.* **45,**956 (1966).
159. J. L. Pope, T. M. Parkinson, and J. A. Olson, *Biochim. Biophys. Acta* **130,** 218 (1966).
160. M. Gracey, V. Burke, and A. Oshim, *Biochim. Biophys. Acta* **225,** 308 (1971).
161. M. Gracey, V. Burke, and A. Oshim, *Scand. J. Gastroenterol.* **6,** 273 (1971).
162. M. Gracey, V. Burke, and C. M. Anderson, *Lancet II,* p. 384 (1969).
163. A. M. Dawson and K. J. Isselbacher, *J. Clin. Invest.* **39,** 730 (1960).
164. J. H. Rosenberg, W. G. Hardison, and D. M. Bull, *N. Engl. J. Med.* **276,** 1391 (1967).
165. R. M. Donaldson, *Fed. Proc.* **26,** 1426 (1967).

166. S. Tabaqchali, J. Hatzioannou, and C. C. Booth, *Lancet* 2, 12 (1968).
167. M. E. Ament, S. S. Shimoda, D. R. Saunders, and C. E. Rubin, *Gastroenterology* 63, 728 (1972).
168. D. M. Small, *Gastroenterology* 52, 607 (1966).
169. M. D. Siperstein, *Curr. Top. Cell Regul.* 2, 74 (1970).
170. J. M. Dietschy and J. D. Wilson, *N. Engl. J. Med.* 282, 1128 (1970).
171. J. M. Dietschy and J. D. Wilson, *N. Engl. J. Med.* 282, 1179 (1970).
172. J. M. Dietschy and J. D. Wilson, *N. Engl. J. Med.* 282, 1241 (1970).
173. E. H. Mosbach, *Arch. Intern. Med.* 130, 478 (1972).
174. G. Salen, G. Nicolau, S. Shefer, and E. H. Mosbach, *Gastroenterology* 69, 676 (1975).
175. Z. R. Vlahcevic, C. C. Bell, J. Buhac, J. T. Farrar, and L. Swell, *Gastroenterology* 59, 165 (1970).
176. Z. R. Vlahcevic, C. C. Bell, and L. Swell, *Gastroenterology* 59, 62 (1970).
177. D. M. Small and S. Rapo, *N. Engl. J. Med.* 283, 53 (1970).
178. T. C. Northfield and A. F. Hofmann, *Gut* 16, 1 (1975).
179. Z. R. Vlahcevic, P. Juttijudata, C. C. Bell, and L. Swell, *Gastroenterology* 62, 1174 (1972).
180. R. H. Dowling, G. D. Bell, and J. White, *Gut* 13, 415 (1972).
181. G. Salen, G. Nicolau and S. Shefer, (abstr.) *Clin. Res.* 21, 523 (1973).
182. G. G. Bonorris, M. J. Coyne, and L. I. Goldstein, (abstr.) *Gastroenterology* 67, 780 (1974).
183. R. D. Adler, L. J. Bennion, W. C. Duane, and S. M. Grundy, *Gastroenterology* 68, 326 (1975).
184. W. G. Linscheer and K. L. Raheja, *Lancet* 2, 551 (1974).
185. W. G. Linscheer, K. L. Raheja, and E. C. Regensburger, *Gastroenterology* 66, 732 (1974).
186. R. N. Redinger and D. M. Small, *J. Clin. Invest.* 52, 161 (1973).
187. J. Lagarriga and J. A. D. Bouchier, *Gut* 14, 956 (1973).
188. J. B. Carey, *Gastroenterology* 41, 285 (1961).
189. D. H. Sandberg, J. Sjövall, K. Sjövall, and T. A. Turner, *J. Lipid Res.* 6, 182 (1965).
190. S. J. Levin and M. K. Schwartz, *Clin. Chem.* 11, 547 (1965).
191. K. Sjövall and J. Sjövall, *Clin. Chem. Acta* 13, 207 (1966).
192. B. Frosch and H. Wagner, *Nature* 213, 404 (1967).
193. G. Neale, B. Lewis, V. Weaver, and D. Panveliwalla, *Gut* 12, 145 (1971).
194. W. Forth, P. Doenecke, and H. Glasner, *Klin. Wochenschr.* 43, 1102 (1965).
195. M. Rauterau, B. Chevrel, and J. Caroli, *Rev. Med.-Chir. Mal. Foie* 42, 167 (1967).
196. J. Makino, S. Nakagawa, and K. Mashimo, *Gastroenterology* 56, 1033 (1969).
197. B. Franz and J. C. Bode, *Klin. Wochenschr.* 52, 522 (1974).
198. I. M. Weiner, J. E. Glasser, and L. Lack, *Am. J. Physiol.* 207, 964 (1964).
199. T. Alström and A. Norman, *Acta Med. Scand.* 191, 521 (1972).
200. P. Back, *Digestion* 10, 322 (1974).
201. A. Norman, B. Strandvik, and R. Zetterström, *Acta Paediatr. Scand.* 58, 59 (1969).
202. A. E. Cowen, M. G. Korman, A. F. Hofmann, and O. W. Cass, *Gastroenterology* 69, 59 (1975).
203. P. Back, *Klin. Wochenschr.* 51, 926 (1973).
204. J. Makino, K. Shinozaki, and S. Nakagawa, *J. Lipid Res.* 15, 132 (1974).
205. A. E. Cowen, M. G. Korman, A. F. Hofmann, O. W. Cass, and S. B. Coffin, *Gastroenterology* 69, 67 (1975).
206. T. S. Low-Beer, M. P. Tyor and L. Lack, *Gastroenterology* 56, 721 (1969).

207. H. L. Sharp and B. L. Mirkin, *Pediatrics* **81**, 116 (1972).
208. P. Berthelot, S. Erlinger, D. Dhumeaux, and A. M. Preaux, *Am. J. Physiol.* **219**, 809 (1970).
209. C. D. Klaassen, *J. Pharmacol. Exp. Ther.* **176**, 743 (1971).
210. K. von Bergmann, H. P. Schwarz, and G. Paumgartner, *Arch. Pharmacol.* **287**, 33 (1975).
211. S. Erlinger, D. Dhumeaux, P. Berthelot, and M. Dumont, *Am. J. Physiol.* **219**, 416 (1970).
212. D. H. Gregory, Z. R. Vlahcevic, P. Schatzki, and L. Swell, *J. Clin. Invest.* **55**, 105 (1975).
213. J. L. Boyer, D. Reno, T. Layden, and J. Schwarz, *J. Clin. Invest.* **53**, 9a (1974).
214. W. J. Simmonds, M. G. Korman, V. L. W. Go, and A. F. Hofmann, *Gastroenterology* **65**, 705 (1973).
215. B. Josephson, *J. Clin. Invest.* **18**, 343 (1939).
216. N. Kaplowitz, M. S. Kok, and N. B. Javitt, *J. Am. Med. Assoc.* **225**, 292 (1973).
217. M. G. Korman, N. F. La Russo, N. E. Hoffman, and A. F. Hofmann, *N. Engl. J. Med.* **292**, 1205 (1975).
218. M. Blum and N. Spitz, *J. Clin. Invest.* **45**, 187 (1966).
219. G. A. Borel and P. Magnenat, *Helv. Med. Acta* **37**, 129 (1973).
220. N. L. La Russo, N. E. Hoffman, A. F. Hofmann, and M. G. Korman, *N. Engl. J. Med.* **292**, 1209 (1975).
221. W. Erb, R. Kröhl, J. Schreiber, and J. Wildgrube, *Z. Gastroenterol.* **10**, 85 (1972).
222. E. Theodor, N. Spitz, and M. H. Slesinger, *Gastroenterology* **55**, 183 (1968).
223. D. Trülzsch, H. Greim, P. Czygan, F. Hutterer, F. Schaffner, H. Popper, D. Y. Cooper, and O. Rosenthal, *Biochemistry* **12**, 76 (1973).
224. K. Einarsson and K. Hellström, *Clin. Sci. Mol. Med.* **46**, 183 (1974).
225. M. D. Kaye, J. E. Struthers, J. S. Tidball, E. DeNiro, and F. Kern, *Clin. Sci. Mol. Med.* **45**, 147 (1973).
226. J. Roovers, E. Evrard, and H. Vanderhaeghe, *Clin. Chim. Acta* **19**, 449 (1968).
227. S. M. Mehta, J. E. Struthers, M. D. Kaye, and J. L. Naylor, *Gastroenterology* **67**, 674 (1974).
228. S. Sherlock, in "Diseases of the Liver and Biliary System," p. 277, Blackwell, Oxford, Edinburgh (1968).
229. V. J. Desmet, in "Progress in Liver Diseases" (H. Popper and F. Schaffner, eds.), p. 97, Grune and Stratton, New York, London (1972).
230. G. P. van Berge Henegouwen, MD-thesis, Catholic University of Nijmegen, The Netherlands (1974).
231. G. P. van Berge Henegouwen and K. H. Brandt, in: "Advances in Bile Acid Research" (S. Matern, J. Hackenschmidt, P. Back, and W. Gerock, eds.), p. 395, Schattauer, Stuttgart–New York (1975).
232. R. H. Dowling, E. Mack, and D. M. Small, *J. Clin. Invest.* **49**, 232 (1970).
233. A. Norman and B. Strandvik, *J. Lab. Clin. Med.* **78**, 181 (1971).
234. D. M. Small, R. H. Dowling, and R. N. Redinger, *Arch. Intern. Med.* **130**, 552 (1972).
235. J. Makino, J. Sjövall, A. Norman, and B. Strandvik, *FEBS Letters* **15**, 161 (1971).
236. A. Norman and B. Strandvik, *Acta Paediatr. Scand.* **62**, 253 (1973).
237. A. Norman, B. Strandvik, and O. Ojamäe, *Acta Paediatr. Scand.* **63**, 97 (1974).
238. P. Back, *Verh. Dtsch. Ges. Inn. Med.* **80**, 404 (1974).

Chapter 5

PATHOPHYSIOLOGY AND DISSOLUTION OF CHOLESTEROL GALLSTONES

Jay W. Marks, George G. Bonorris, and Leslie J. Schoenfield

Section of Gastroenterology
Department of Medicine
Cedars-Sinai Medical Center
Los Angeles, California

Bile is a complex fluid containing three lipids—cholesterol, bile acids, and phospholipid—as well as conjugated bilirubin, trace amounts of protein, and inorganic ions. Normally these substances are fully solubilized in bile. Under abnormal conditions, however, some of these constituents may become insoluble and precipitate. If further precipitation occurs and the precipitated complexes grow in size, gallstones will result.

Cholesterol, bilirubin, and calcium carbonate gallstones all occur in man, while lithocholic acid gallstones can be produced experimentally in animals (1,2). Cholesterol gallstones, either pure or mixed (containing more than 70% cholesterol) account for approximately 85% of gallstones in the United States with bilirubin pigment gallstones accounting for the remaining 15% (3).

An estimated 16 million Americans are afflicted with gallstones (4). Because of the significance of gallstones as a national health problem, much effort has been expended on cholesterol gallstone research. The purpose of this chapter is to review research contributing to our present understanding of cholesterol gallstone pathophysiology and its consequence, experimental gallstone dissolution.

* Supported in part by NIH Grant AM 15631.

I. BILIARY LIPID PHYSIOLOGY

The pathophysiological mechanisms that are relevant to cholesterol gallstone formation involve the major lipid constituents of bile—cholesterol, bile acids, and phospholipid. All three lipids similarly undergo the processes of hepatic synthesis, biliary secretion, intestinal transformation, and reabsorption. Whereas cholesterol and phospholipid are important in the physiology of all cells, bile acids function primarily within the enterohepatic circulation.

A. Cholesterol

1. Metabolism

De novo synthesis is the major source of body cholesterol in man and lower animals, whereas dietary cholesterol provides a quantitatively less important and dispensable source. The ability among tissues to synthesize cholesterol, however, varies greatly (5). Extensive *in vitro* studies with rat and monkey tissue slices have shown that liver and intestine have by far the greatest capacity for cholesterol synthesis. Since these tissues also are those most influenced by the primary regulators of cholesterol synthesis, it has been proposed that the liver and intestine are predominantly responsible for the maintenance of cholesterol homeostasis. Less extensive data with human tissue suggest that a similar situation exists in man.

The cholesterol synthetic pathway has been most extensively studied in the hepatocyte where the enzymes involved in synthesis are found in both the microsomal and soluble fractions (6). Three acetyl coenzyme A fragments undergo condensation to form β-hydroxy-β-methylglutaryl-coenzyme A (HMG-CoA) which is then reduced to form mevalonate. When three mevalonate molecules undergo phosphorylation, decarboxylation, and condensation farnesyl pyrophosphate is formed. Two farnesyl pyrophosphate molecules then condense to form squalene. Cyclization of squalene to lanosterol is followed by various nuclear changes that ultimately result in the formation of cholesterol. The enzyme responsible for the reduction of HMG-CoA to mevalonate, HMG-CoA reductase, is rate limiting (7).

Unlike the majority of plasma and hepatic cholesterol, which is esterified at the 3-position with fatty acid, biliary cholesterol is virtually all unesterified. In the intestine biliary free cholesterol mixes with esterified cholesterol from other sources, including the diet, intestinal secretions, and desquamated intestinal mucosal cells. Intestinal cholesterol may exist in different pools since it has been found in the rat that endogenously synthesized cholesterol is less efficiently absorbed than exogenous cholesterol (8). The esterified cholesterol in the intestinal lumen is hydrolyzed by pancreatic cholesterol esterase, and the resultant enlarged pool of free cholesterol is partially absorbed, mixing with the intracellular pool of the intestinal epithelial cell. Much of the cholesterol is reesterified, released into the intestinal lymph as chylomicrons, and returned to the liver. The cholesterol that is not absorbed is excreted as fecal neutral steroid.

2. Effect of Diet

Bile acids through their effect on hepatic and intestinal cholesterol synthesis and on cholesterol absorption are endogenous regulators of cholesterol metabolism (9). The two major exogenous regulators of cholesterol metabolism are dietary cholesterol and caloric intake. It is important to stress that cholesterol absorption in man is incomplete and limited. Man absorbs approximately 40% of the cholesterol in his normal diet (10,11). Within the normal range of cholesterol intake cholesterol absorption is linear, but if cholesterol intake exceeds this range, cholesterol absorption does not increase proportionally (12). For example, on an extremely high cholesterol intake of 2900 mg/day, only 12% or 340 mg/day is absorbed. It has been suggested that the maximal absorption is actually somewhat higher, 600 mg/day (12), perhaps even 1 g/day (13). The ability to absorb dietary cholesterol also appears to vary among individuals (13).

Undoubtedly, the limited ability of man to absorb dietary cholesterol is an important protective mechanism and accounts for the less than 25% increase in plasma cholesterol seen with a high cholesterol intake. In contrast, animals such as the rabbit, whose capacity for cholesterol absorption is much greater, become grossly hyperlipemic with cholesterol feeding (14). There are two metabolic implications of this phenomenon in man (11). First, any increase in cholesterol absorption can be completely compensated for by feedback inhibition of cholesterol synthesis. Second, endogenous cholesterol synthesis, approximately 1.1 g/day, is rarely totally suppressed even on very high cholesterol diets (15).

Total body cholesterol synthesis in man is inhibited by a very high cholesterol intake though the degree of inhibition varies from individual to individual (13). In monkey liver tissue slices, high dietary cholesterol depresses cholesterol synthetic activity, and a similar depression has been found to occur in liver biopsy tissue from man (16). The effect of dietary cholesterol on intestinal synthesis, however, is less than that on liver synthesis in monkey tissue slices. Moreover, it has been suggested that in man, the inhibitory effect of dietary cholesterol on intestinal cholesterol synthesis is actually insignificant (5). Thus, under conditions of high cholesterol intake, the intestinal mucosa is the predominant source of cholesterol.

Starvation reduces the cholesterol synthetic capacity of monkey liver slices (5). Similarly, in obese persons undergoing weight reduction, cholesterol synthesis decreases (17). However, the decrease cannot be attributed solely to calorie restriction since low calorie diets are also low in fat content. Dietary fat may be necessary for cholesterol absorption (13) and is known to affect serum cholesterol (18). These dietary triglycerides, however, have little effect upon the actual synthesis of cholesterol (18,19).

3. Body Pools

Body cholesterol is conceptualized as existing in three pools (described as rapidly, slowly, and very slowly turning over) which are in dynamic equilibrium (5). Plasma, hepatic, and biliary cholesterol are all part of the rapidly turning over pool. Recent evidence suggests, however, that in the liver there is a microsomal subpool of cholesterol destined for bile and that cholesterol recently synthesized by the liver may be preferentially excreted in the bile (20,21).

Since biliary cholesterol is in equilibrium with other cholesterol pools, one might expect to find a correlation between biliary cholesterol secretion and cholesterol intake or absorption. Indeed, cholesterol feeding, despite inhibiting hepatic cholesterol synthesis, increases biliary cholesterol secretion (22) and fecal excretion of endogenous neutral steroids (13). Also, in the fasting rhesus monkey, biliary cholesterol secretion is low (23), though there is conflicting data regarding this point (24). In man, there is evidence that during weight reduction, cholesterol mobilized from adipose tissue is excreted in bile (25). Thus, changes in biliary cholesterol secretion, in at least some instances, may reflect changes in overall cholesterol metabolism. The relationship is complex, however, and exceptions exist. For example, some patients with type II hypercholesterolemia do not develop an increase in biliary cholesterol secretion until very large amounts of cholesterol are fed (22). Accordingly it has been suggested that their threshold for biliary cholesterol secretion is high.

B. Bile Acids

1. Synthesis

Synthesis of the two primary bile acids, cholic acid and chenodeoxycholic acid from their precursor, cholesterol, involves nuclear changes and side chain degradation of the cholesterol molecule (6,26). Synthesis of cholic acid, a trihydroxy bile acid, is initiated in the hepatic microsomes with 7α-hydroxylation of cholesterol by the enzyme 7α-hydroxylase. Further chemical modifications of the nucleus taking place in the microsomes and cytosol include epimerization of the hydroxyl group in the 3-position to the α-configuration, saturation of the Δ^5-double bond, and introduction of a hydroxyl group at the 12α-position. Cholesterol side chain degradation is initiated with microsomal 26-hydroxylation and is followed by several oxidative steps that occur in the microsomes, cytosol, and mitochondria. A lysosomal acyl transferase then catalyzes the conjugation of the product, cholic acid, with taurine or glycine, and it is these conjugates that are secreted into the biliary tract. Similar steps are involved in the synthesis of chenodeoxycholic acid, a dihydroxy bile acid, but the pathway is not as well defined as that for cholic acid. The 7α-hydroxylation step is rate-limiting for the synthesis of both bile acids. In man, about 60% of lithocholic acid, a secondary bile acid, is sulfated (27,28).

2. Enterohepatic Circulation

The conjugated bile acids secreted by the liver are stored in the gallbladder and are then periodically discharged into the intestine. Most of the intestinal conjugated bile acids are actively absorbed in the terminal ileum although small amounts are passively absorbed along the length of the intestine (29). Less than 20% of the intestinal conjugated bile acids are subjected to enzymatic deconjugation by bacteria in the terminal ileum and colon (30). The unconjugated bile acids thus formed are actively reabsorbed in the ileum or passively reabsorbed in the ileum and colon. Unconjugated cholic and chenodeoxycholic acids are enzymatically 7α-dehydroxylated by bacteria primarily in the colon to deoxycholic and lithocholic acids, their respective di- and monohydroxy secondary bile acids, which are partially absorbed by passive diffusion. Thus, human bile may contain up to 35% deoxycholic and 4% lithocholic acids.

Ninety-five percent of intestinal bile acids are reabsorbed in normal man; only 5% are excreted as fecal acidic steroids. Thus a small bile acid pool is recycled 6–10 times a day via an efficient portal enterohepatic circulation to provide large amounts of bile acids for intestinal fat solubilization and absorption (31).

3. Regulation of Synthesis

Regulation of bile acid synthesis depends on bile acid excretion, cholesterol balance, and caloric intake. Under normal conditions in man, hepatic bile acid synthesis amounts to 200–500 mg/day, sufficient to replace fecal loss and maintain a pool size of approximately 3.0 g (32). If bile acid excretion is increased by ileal exclusion or oral administration of bile acid binding resin, synthesis of bile acids increases (33). If bile acids are fed to humans, bile acid synthesis decreases (9). The site of this feedback inhibition of bile acid synthesis appears to be at the first and rate-limiting step of bile acid synthesis, 7α-hydroxylation, since biliary diversion increases 7α-hydroxylase activity and bile acid feeding decreases it (34).

The precise role of cholesterol in the regulation of bile acid synthesis has not been defined. In dogs, dietary supplementation with cholesterol results in an enlarged taurocholic acid pool with a decreased half-life of the bile acid (35). In man, dietary cholesterol supplementation may lead to increased biliary bile acid secretion (22) and fecal bile acid excretion (11), although this has been disputed (13). Polyunsaturated fats, which mobilize tissue cholesterol, cause increased fecal neutral and acidic steroid excretion in some patients, suggesting increased conversion of cholesterol to bile acids (18,36). Hepatic synthesis of bile acids, therefore, may be stimulated by cholesterol. Caloric and triglyceride intake must also be important in the regulation of bile acid synthesis, for in the rhesus monkey fasting decreases while triglycerides increase both bile acid synthesis and pool size (23).

C. Phospholipids

More than 90% of the phospholipid in human bile is lecithin (37), primarily of the linoleyl–palmityl type (38). *De novo* synthesis of biliary lecithin occurs in the liver predominantly by transfer of choline to diglyceride in the microsome, although it has been shown in rat liver (39), and perhaps human liver, (37) that methylation of ethanolamine phosphoglycerides is also a significant pathway for lecithin synthesis. Three to six grams per day of newly synthesized biliary lecithin, accounting for the majority of hepatic phospholipid synthesis, is preferentially excreted into bile but is also in equilibrium with hepatic and plasma lecithin pools (37,40).

Lecithin within the intestine is either exogenous (dietary) or endogenous in origin with biliary lecithin contributing approximately half of the endogenous portion (41). In man, intestinal lecithin is hydrolyzed by pancreatic phospholipases to lysolecithin, absorbed, reacylated, and secreted into the lymphatics as chylomicron lecithin (42). The enterohepatic circulation of intact lecithin, therefore, is insignificant (43). Moreover, it is disputed whether large oral supplements of lecithin cause increased biliary lecithin secretion (44–46).

Bile acids in the enterohepatic circulation are the most important regulators of hepatic synthesis and biliary excretion of lecithin. Availability of the major substrates of lecithin synthesis—glycerol, glucose, and fatty acids—does not appear to be rate-limiting. There is no evidence for or against feedback control of phospholipid synthesis. Changes in dietary fat are known to cause alterations in the types of lecithin synthesized by the liver (47,48), and in *in vitro* studies with human liver slices, changes in saturation of fatty acid substrate altered the synthetic rates of choline-phosphoglyceride, a precursor of biliary lecithin (49). Finally, in the rhesus monkey, fasting reduces while triglycerides increase biliary phospholipid secretion (23).

D. Bile Acid Regulation of Cholesterol and Phospholipid Metabolism

The metabolism of the three major biliary lipids are complexly interrelated. As already mentioned, cholesterol, triglyceride, and caloric intake are important exogenous regulators of cholesterol, phospholipid, and bile acid synthesis. In addition, bile acids play a major regulatory role in cholesterol and phospholipid metabolism.

Bile acids are indispensable to the absorption of dietary cholesterol, for absence of luminal bile acids is associated with a marked decrease in absorption of exogenous (50) and endogenous (36) cholesterol. Also, addition of the trihydroxy bile acid (cholic acid), but not di- or monohydroxy bile acids, to a diet containing cholesterol, increases cholesterol absorption in rats (51,52). The mechanisms involved are undoubtedly multiple and may include interaction of the bile acid with cholesterol ester hydrolase in the intestinal lumen, uptake of cholesterol into the intestinal cell, intramucosal transport and esterification of cholesterol, and chylomicron formation (53).

Bile acids in the enterohepatic circulation regulate intestinal and hepatic cholesterol synthesis. Thus, the striking increase in intestinal cholesterol synthesis that occurs in rats after diversion of bile from the intestinal lumen (54) can be prevented by the intraluminal infusion of pure bile acids (54,55). Furthermore, in the rat, ingestion of a bile acid binding resin leads to increased hepatic cholesterol synthesis (56) while supraphysiological amounts of oral bile acids result in inhibition of hepatic cholesterol synthesis (57). Thus, in animals as well as in man (36), bile acids regulate cholesterol synthesis in the intestine and in the liver. Whether this hepatic regulation occurs by a direct or an indirect mechanism through cholesterol absorption, however, is not clear (5,53).

Bile acids regulate hepatic phospholipid synthesis and secretion. Addition of conjugated cholic acid to human liver tissue slices results in increased choline-phosphoglyceride synthesis (49). Also, in both the perfused rat liver and bile fistula rat, conjugated cholic acid causes an increase in hepatic lecithin synthesis and biliary secretion (40,58).

E. Biliary Secretion

1. Hepatic Uptake of Bile Acids

The secretion of biliary lipids and the flow of bile are largely dependent on bile acids, though factors independent of bile acids also appear to be important. Bile acids returning to the liver via the enterohepatic circulation account for 95% of biliary bile acid secretion. The remainder is accounted for by *de novo* synthesis which is under negative feedback control as previously discussed. Because of the quantitative significance of the returning bile acids, hepatic uptake of bile acids can be considered the first step in bile acid secretion.

Primary and secondary bile acids returning via the portal vein are removed from the blood by the hepatocyte, transported through the cell, and secreted into the bile canaliculus. Removal of bile acids from blood is very efficient, with greater than 90% removal in just one passage through the liver. The capacity for secretion, though saturable, is high. Therefore, under normal conditions neither uptake nor secretion is likely to limit the flow of bile acids into the biliary tract.

2. Lipids

Bile acid secretion is a major regulator of cholesterol and lecithin secretion (59). In man the relationship between bile acid and cholesterol secretion rates is linear, while the relationship between bile acid and lecithin secretion rates is nonlinear (Fig. 1). Thus, decreasing bile acid secretion is associated with decreasing cholesterol and lecithin secretion, but the decrease in lecithin is greater than the decrease in cholesterol. At very low bile acid secretory rates cholesterol secretion remains relatively high, suggesting significant bile acid independent cholesterol secretion. Lecithin secretion appears to be limited and completely dependent on bile acid secretion, although in the rhesus monkey, bile acid independent lecithin secretion can be demonstrated with severe, prolonged interruption of the enterohepatic circulation (60).

During periods of decreased return of bile acids to the liver, for example with fasting or interruption of the enterohepatic circulations, bile flow will be low, bile cholesterol content will be relatively high, and bile acid and lecithin content relatively low. Thus, biliary lipid composition will

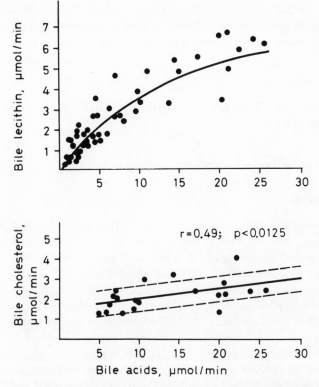

Fig. 1. Relationship between the biliary secretion rate of bile acids and the secretion rate of lecithin and cholesterol. Taken from Scherstén (59).

normally vary throughout the day with relatively more cholesterol during periods of fasting.

There are two ways in which bile acids might effect increased cholesterol and lecithin secretion. Bile acids could primarily increase the synthesis of these lipids with increased secretion occurring as a secondary phenomenon. Alternatively, the influence of bile acids might be directly on lipid secretion with increased synthesis a secondary phenomenon. There is scant evidence to support the theory that biliary cholesterol and phospholipid are derived from the canalicular membrane (61).

3. Water and Electrolytes

Bile acids have a stimulatory effect on bile flow (62). The increase in flow is caused by net water secretion into the biliary tract with proportional increases in Na$^+$, K$^+$, and Cl$^-$ secretion (63). The mechanism of this stimulation is unknown but appears to involve more than the osmotic effect of the bile acids. There is also a significant bile acid independent process affecting bile flow which probably involves active sodium transport (64).

II. PATHOPHYSIOLOGY OF CHOLESTEROL GALLSTONE FORMATION

A. Stages of Gallstone Formation

1. *Saturation*

Cholesterol gallstone formation is conveniently divided into three stages: saturation, nucleation, and growth. This should not imply that each stage occurs independently of the other two since all three can often be seen at the same time in a gallstone-containing gallbladder.

The mechanism of cholesterol solubility in bile is complex—for bile is an aqueous solution and cholesterol is water-insoluble. Bile acids are amphipathic, that is, they contain both hydrophilic polar groups and hydrophobic aliphatic or aromatic portions. This amphipathic nature of bile acids is responsible for their forming simple micelles in aqueous solutions. A micelle may be described as a colloidal aggregation of an amphipathic compound or detergent in which the hydrophobic portion faces inward and the hydrophilic groups point outward. Once the bile acid concentration rises above a certain level—the critical micellar concentration (CMC)—bile acid micelles form spontaneously. Micelle formation is influenced not only by the concentrations of bile acids, but also by the concentration of biliary solids, the molecular structure of the bile acid (conjugation and hydroxyl group position), temperature, pH, and counterion concentration.

Bile acid micelles have only a slight capacity for solubilization of cholesterol, but they are capable of solubilizing and incorporating lecithin with a lecithin:bile acid molar ratio of 2:1. Lecithin, though water-insoluble itself, allows water to penetrate its crystalline structure and causes the micelle, now referred to as a mixed micelle, to swell. The lecithin fraction of the mixed micelle is also capable of solubilizing cholesterol in equimolar amounts. There are two important consequences of bile acid–lecithin, mixed micelle formation. First, the micelle develops a greater capacity to solubilize cholesterol. Second, the CMC is reduced, allowing solubilization of cholesterol to occur at lower bile acid concentrations. Efficient cholesterol solubilization in bile, therefore, depends on the presence of both bile acids and lecithin (65).

Cholesterol solubility in bile is nevertheless limited. When the maximal solubility for cholesterol is reached or exceeded, that is, when bile becomes saturated or supersaturated, the first requirement for gallstone formation has been met. The limits of cholesterol solubility can be determined *in vitro* and used to distinguish saturated and supersaturated from unsaturated bile. Under any standard conditions of total biliary solids—concentration, temperature, pH, and counterion concentration—cholesterol solubility will depend on the relative concentrations of cholesterol, bile acids, and lecithin. This is graphically illustrated by the use of triangular coordinates wherein the axes represent the percent of the total moles of bile acids, phospholipids (lecithin), and cholesterol constituted by each of these components (Fig. 2). Any bile sample, therefore, can be represented by a single point on such coordinates. By varying the relative concentrations of cholesterol, bile acids, and lecithin *in vitro*, Admirand and Small (66) defined the upper limit for biliary cholesterol solubility on triangular coordinates.

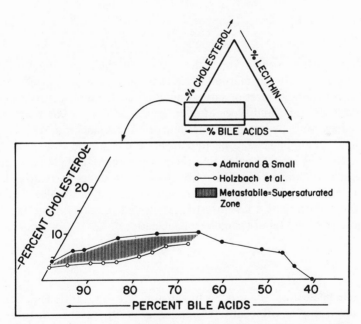

Fig. 2. The relative molar proportions of biliary lipids plotted on triangular coordinates. The axes represent the percent of the total moles of bile acids, lecithin, and cholesterol constituted by each of these components which total 100 percent. The limits of cholesterol solubility are shown as defined by Admirand and Small (66) and Holzbach *et al.* (67) with the metastable supersaturated zone between them.

Holzbach *et al.* (67) redefined the limit of cholesterol solubility and described a metastable supersaturated zone (68). Points below the redefined limit represent bile that is unsaturated with cholesterol while points within the zone of metastable supersaturation represent bile in which cholesterol exists in micellar, liquid crystalline, and crystalline phases. The liquid crystals are unstable and their cholesterol may either return to the micelle or precipitate as crystals. Formation of cholesterol crystals occurs slowly at the lower limit of the zone, where it may take hours or even days for cholesterol nucleation and precipitation to occur (69), and rapidly at the upper limit. Above the zone, cholesterol also exists in three phases.

Methods other than triangular coordinates have been used to express cholesterol solubility in bile. These have included the ratio of the concentrations of cholesterol to that of bile acids plus lecithin and the lithogenic index or saturation index which is the ratio of the actual amount of cholesterol present in a bile specimen to the maximum amount of cholesterol that can be dissolved in that specimen (70). The indices can be determined from either the triangular coordinate diagram or a formula (71).

2. Nucleation

Very little is known about the process of cholesterol nucleation. It is probably not a simple matter of cholesterol crystallization whenever bile reaches a threshold level of saturation. Physicochemical factors such as bile acid structure, pH, and types of counterions, as well as factors responsible for seeding, such as bile pigments, mucoproteins, bacteria, or precipitated inorganic compounds may be important. For example, studies of cholesterol gallstones have revealed that their centers contain significant amounts of pigment and mucus (72–75). Also, an excellent direct correlation between the presence of biliary pigment precipitates and cholesterol gallstones has been noted (76). Only a few investigators have begun to explore factors that promote or inhibit biliary cholesterol nucleation (77,78).

3. Growth

Like nucleation, the process of gallstone growth is little understood and probably complex. Recently Osuga *et al.* (79) have made a scanning electron microscope study of gallstones and bile sediments from gallstone patients and have found that the basic units of both micro- and macroscopic stones are laminated cholesterol crystals (Figs. 3 and 4). These laminated crystal units aggregate randomly, radially, or concentrically and show evidence of dissolution, implying that gallstones result from net cholesterol

Fig. 3. Scanning electron photomicrograph of a cluster of laminated cholesterol crystals aggregated at random. Taken from Osuga *et al.* (79).

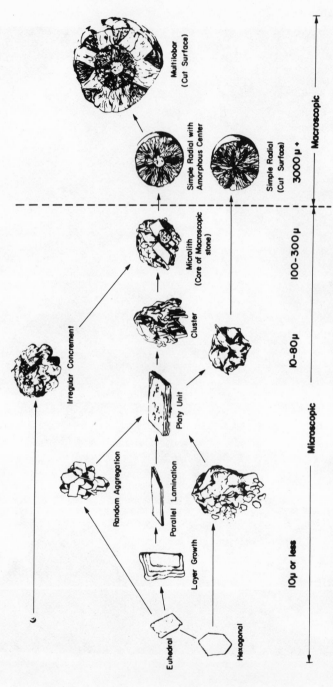

Fig. 4. A diagram of the different ways in which macroscopic gallstones are probably formed from microscopic forms in the common duct and gall-bladder bile of man. Taken from Osuga *et al.* (79).

precipitation in bile where conditions alternately favor precipitation or dissolution. Although some cholesterol gallstones are virtually pure cholesterol, most in the United States are mixed—containing cholesterol, pigment, and calcium salts. Cholesterol gallstones may also contain varying small amounts of bile pigments, fatty acids, protein, and phospholipids (3). Are the noncholesterol constituents entrapped bystanders or are they active participants in gallstone formation and growth? Does gallbladder bile composition change during the growth of some stones (80)? Are such changes coincidental or contributory? The answers to these questions are unknown.

B. Mechanism of Cholesterol Saturation

Precipitation of cholesterol is dependent on saturation of bile with cholesterol. There is a good direct correlation between biliary cholesterol saturation and the prevalence of cholesterol gallstones in different human population groups and animal species (81). Also, there is indirect evidence that the abnormalities of biliary lipid metabolism seen in some populations of gallstone patients precede the development of gallstones (82). Thus, physicochemical and epidemiological considerations strongly suggest that saturated bile is a prerequisite of cholesterol gallstone formation. This does not imply that cholesterol saturation is the only necessary condition, but it has been the most extensively studied factor in gallstone formation.

1. Source of Saturated Bile

Bile from cholesterol gallstone patients is more frequently saturated with cholesterol than bile from normals, although there is overlap between the two groups (67). Although it is clear that the liver is the source of the saturated bile (25,83), it is not known whether the basic abnormality is in the liver or with the other factors regulating biliary lipid composition (primarily bile acid secretion) or both. Just as the return of bile acids to the liver varies during the day, biliary cholesterol saturation will also normally vary throughout the day, reaching supersaturated levels even in some normal individuals. Since even normal persons are capable of secreting bile supersaturated with cholesterol during overnight fasting, the only difference between gallstone patients and normal persons may be that gallstone patients secrete saturated bile for a larger part of each day. Although a short-term secretion study supports this possibility (84), a study in which secretion has been measured over a 24-hour period does not do so (85).

There are three ways in which hepatic bile may become saturated with cholesterol. Cholesterol secretion may be increased relative to bile acid and phospholipid secretion. Conversely, bile acid or phospholipid secretion may be low relative to cholesterol secretion.

2. Reduced Bile Acid Pool

It is well documented that the bile acid pool in cholesterol gallstone patients is abnormally small (86,87), either as a primary or secondary phenomenon. The small pool size might lead to decreased bile acid secretion and increased biliary cholesterol saturation. There are three possible reasons for the decreased pool of bile acids—increased loss, decreased synthesis, or increased cycling frequency.

Normally, loss of bile acids occurs solely through the gut as a consequence of incomplete absorption and bacterial transformation. An increase in fecal bile acid excretion, possibly due to an increased bile acid recycling frequency or decreased efficiency of reabsorption, might lead to decreased bile acid pool size and hepatic secretion. Patients with ileal dysfunction as a result of disease or surgery have a high incidence of cholesterol gallstones (88) associated with increased fecal bile acid excretion (89), a decreased bile acid pool (90), and bile that is supersaturated with cholesterol (91). Furthermore, the centers of human cholesterol gallstones contain a lower than normal proportion of secondary bile acids and taurine conjugates (92), and bile with a similar composition is found in patients with ileal dysfunction (93). This suggests that there is interruption of the enterohepatic circulation during early cholesterol gallstone formation.

There is a positive correlation between decreasing cholic acid half-life and decreasing cholic acid pool size in cholesterol gallstone patients, though a similar correlation for chenodeoxycholic acid does not exist (94). If increased bile acid loss is occurring, it should be possible to demonstrate decreased bile acid half-life and increased bile acid synthesis. In gallstone patients, investigators have found the half-lives of cholic and chenodeoxycholic acid to be decreased (82) or normal (85,86), but their synthesis has been decreased (86,95) or normal (85,95).

It would take a rather large loss of bile acids, at least 20% of hepatic secretion, to affect the bile acid pool if feedback regulation was intact. Increased fecal excretion of bile acids, however, has not been demonstrated in cholesterol gallstone patients. Furthermore, it has been shown in the rhesus monkey that even chronic, severe interruption of the enterohepatic circulation increase biliary cholesterol saturation only transiently (60). Thus, there is inadequate evidence to implicate increased bile acid loss in cholesterol gallstone formation, and it would appear that if bile acid loss is increased, another factor such as abnormal feedback regulation must also be important.

Another possible explanation for the decreased bile acid pool size is that bile acid synthesis is abnormally low. In one study low cholic acid synthesis was demonstrated (86), and in another, low chenodeoxycholic acid synthesis (95), though the latter could not be confirmed (85). Still another study found normal taurocholate synthesis despite a decrease in taurocholate half-life (87). It can be argued that the presence of a small bile acid pool size and a minimal if any increase in fecal bile acid excretion is evidence of inappropriately low bile acid synthesis. This argument would only hold if the return of bile acids to the liver was low and an appropriate increase in synthesis did not occur. It is conceivable that despite a small bile acid pool, return of bile acids to the liver is normal because of an increased recycling frequency (85).

Inappropriately low synthesis and secretion of bile acids could be due to abnormalities in the regulatory or synthetic functions of the liver. There is little data relating to the integrity of the regulatory mechanisms for bile acid synthesis, although Banfield and Admirand have presented preliminary evidence that gallstone patients respond to cholestyramine-induced bile acid depletion with increased bile acid synthesis and maintain their pretreatment bile acid pool size (96). Vlahcevic and coworkers demonstrated that cholecystectomized cholesterol gallstone patients have an increased turnover rate of bile acids that is not accompanied by increased bile acid synthesis (97). They hypothesize that increased bile acid loss and insensitive feedback regulation leading to inadequate bile acid synthesis and secretion is the basic abnormality in cholesterol gallstone formation. As for the synthetic mechanism itself, Grundy *et al.* have suggested that the basic defect is inadequate conversion of cholesterol into bile acids (98). Two groups have indeed shown, *in vitro,* a reduced level of hepatic cholesterol 7α-hydroxylase activity in cholesterol gallstone patients (99,100).

Inappropriate bile acid synthesis might result in a small bile acid pool which in turn might result in low bile acid secretion and increased biliary cholesterol saturation. The presence of a small bile acid pool, however, does not necessarily mean that there is decreased biliary secretion of bile acids. Whereas one study suggests that bile acid secretion is low in cholesterol gallstone patients (84), another shows no difference from normals (85). Low-Beer and Pomare (94) and Northfield and Hofmann (85) have suggested that bile acid secretion is normal in cholesterol gallstone patients even with a decreased bile acid pool because of an increased bile acid recycling frequency.

It is possible that the small bile acid pool in gallstone patients is not related to inappropriate synthesis. According to one hypothesis, impaired gallbladder storage or enhanced gallbladder emptying causes increased recycling frequency of the bile acid pool (Fig. 5). The frequent recycling

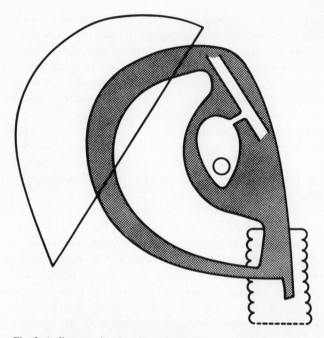

Fig. 5. A diagram showing the enterohepatic circulation of the bile acid pool (stippled). The quality of bile acids returning to the liver and the pool size are dependent upon hepatic formation and secretion, gallbladder function, and intestinal absorption (101).

inhibits bile acid synthesis and results in a decreased bile acid pool (85,87). During periods when most of the bile acid pool is in the gallbladder as, for example, during overnight fasting, cholesterol from supersaturated hepatic bile may overwhelm the cholesterol solubilizing capacity of the small bile acid pool within the gallbladder. The basic abnormality would be one of abnormal gallbladder function (rather than hepatic synthesis) leading to an increased recycling frequency and secondarily to a small bile acid pool.

3. Increased Cholesterol Secretion

Supersaturation of bile with cholesterol might be due to increased cholesterol secretion. Grundy *et al.* found a greater cholesterol output in gallstone patients than normal controls but this was apparently related to obesity (84). Obesity thus appears to be a significant factor in the formation of supersaturated bile.

The activity of hepatic HMG-CoA reductase, the rate limiting enzyme in cholesterol synthesis, has been found to be increased in cholesterol gallstone patients (100,102), and in conjunction with the decrease in 7α-hydroxylase activity, might be responsible for a relative increase in cholesterol synthesis as compared with bile acid synthesis. If enzyme activity reflects synthesis and secretion, cholesterol gallstone patients could thus have bile more saturated with cholesterol than normals even if bile acid return to the liver was identical in the two groups.

4. Phospholipid Secretion

The third possible cause of bile supersaturated with cholesterol is decreased secretion of biliary phospholipids. Most investigators have found the relative proportion of phospholipids in the bile of cholesterol gallstone patients to be normal (95,103). Therefore, it does not appear that abnormalities in phospholipid metabolism alone are responsible for cholesterol saturation. While decreased phospholipid secretion might occur as a result of decreased bile acid secretion, phospholipid secretion as directly measured in cholesterol gallstone patients is normal (84,85).

C. Role of the Gallbladder

The gallbladder, either primarily or secondarily, must be important in the pathogenesis of cholesterol gallstones since most gallstones form in the gallbladder, and most patients do not have recurrent cholesterol gallstones following cholecystectomy.

1. Effects of Cholecystectomy

There is some confusion as to the effects of cholecystectomy on biliary lipid composition. While some investigators have found reversion of saturated bile to normal following cholecystectomy (104,105), others have found no change (97,106). Even if lipid composition returns to normal following cholecystectomy, it would be hazardous to implicate the gallbladder as a primary etiological factor in the production of saturated bile since the normal physiological consequences of cholecystectomy, such as increased bile acid recycling, might be responsible for the desaturation (24,107).

2. Reservoir Function

The gallbladder might simply serve as a reservoir wherein nucleation and growth of gallstones occurs. Northfield and Hofmann (85), as previously discussed, speculated that gallbladder bile might become supersaturated because of the small bile acid pool. If cholesterol crystallization *in vivo* is time dependent as it is *in vitro,* storage of bile in the gallbladder would promote cholesterol precipitation. Stratification of bile in the gallbladder, a naturally occurring phenomenon, would further increase the probability of cholesterol precipitation (108). Also, Osuga *et al.* (79) have stressed that conditions in the gallbladder—progressive concentration, constant temperature, occasional mixing, and abundance of mucus—are ideal for crystallization. Finally, the role of decreased emptying of the gallbladder in the production of saturated bile or cholesterol nucleation remains hypothetical, though under some conditions, such as pregnancy, gallbladder emptying may be impaired (109).

3. Lipid Reabsorption

Absorption of lecithin or bile acids by the gallbladder could give rise to cholesterol saturated bile. Experiments in a guinea pig model, however, have shown that the normal gallbladder is only capable of absorbing lecithin and bile acids at a rate of 3%/h (110,111). Acute inflammation of the gallbladder, as would occur in acute cholecystitis, increases the absorption rate of bile acids, but chronic cholecystitis decreases it (112). In acute cholecystitis, lecithin may be hydrolyzed to lysolecithin and oleic acid by biliary phospholipase A with rapid absorption of the oleic acid (113). The lysolecithin, which is itself more rapidly absorbed than lecithin, is also less efficient than lecithin in cholesterol solubilization. Arguing against a role for absorption in the pathogenesis of cholesterol gallstones is the fact that cholesterol saturation of hepatic bile is similar to or greater than gallbladder bile in gallstone patients. The evidence, therefore, suggests that augmented gallbladder absorption may contribute to biliary cholesterol saturation in acute cholecystitis but has no role in the initial or chronic stages of gallstone disease.

D. Predisposing Factors

1. Heredity

The role of heredity in cholesterol gallstone formation in Caucasians has been considered by several investigators (114,115). In familial aggregation studies heredity appears to be a predisposing factor, but in twin studies its significance is less striking. Gallstones are probably of multifactorial causation and result from the interaction of environmental influences and the individual's genetic predisposition.

The incidence of cholesterol gallstones varies markedly among racial groups. One of the most carefully studied populations is the North American Indian in whom the incidence of cholesterol gallstones is very high: 70% in women over age 30 and men over age 55 (116). The pathogenesis of gallstones in this group is believed to be basically the same as in Caucasians (98). No environmental factor has been identified that might be responsible for this tremendous incidence, and although an environmental factor is not completely excluded, it is probable that North American Indians are genetically predisposed to cholesterol gallstones.

2. Sex and Sex Hormones

The higher incidence and earlier onset of gallstone disease in women as compared with men is well known. With or without gallstones biliary lipid composition may be different in males and females. In one study, normal women were found to have significantly greater total biliary lipids and different biliary bile acid ratios than men (117), but the relevance of these observations is not clear.

Pregnancy is associated with an impaired hormonal mechanism for gallbladder contraction (109), and this may be related to the observation that Caucasian women with greater numbers of pregnancies develop gallstones more frequently than those who have fewer pregnancies (116). Similarly, women receiving oral contraceptives or postmenopausal estrogen therapy have an incidence of surgically confirmed cholelithiasis that is twice the normal incidence (118,119). It has been demonstrated that combination oral contraceptives increase biliary cholesterol saturation and suggested that decreased bile acid synthesis may be responsible (124).

3. Age

In both Caucasians and North American Indians an increasing prevalence of gallstones is associated with increasing age (116,120). There is, however, no data correlating biliary lipid changes with age in either Caucasians or Indians. An alternative explanation for this association is that it may take many years for gallstones to form in saturated bile.

4. Obesity

Obese persons have increased synthesis, biliary secretion, and fecal excretion of cholesterol when compared with normals (17,84). In obese persons, during caloric restriction and weight reduction, even without fasting, biliary cholesterol saturation may be increased (25). Thus, dieting may increase the risk of cholesterol gallstones. After weight reduction, there are decreases in cholesterol synthesis, biliary secretion and saturation, and fecal excretion (17). In Caucasians gallstone prevalence is correlated with body weight.

5. Diet

The possible influence of cholesterol intake on biliary lipid metabolism has already been discussed. Grundy (36) and Connor *et al.* (18) have studied the effects of polyunsaturated fats on lipid metabolism. It appears from their work that the effects of polyunsaturates are multiple and that there is heterogeneity among individuals with regard to response. A diet high in polyunsaturated fat causes increased fecal acidic and/or neutral steroid excretion (presumably by mobilization of adipose tissue cholesterol), but this does not occur in all normal or hyperlipidemic persons. The variation in response may be related in part to basic alterations of lipid metabolism in the hyperlipidemias, but unknown factors are at least as significant. Impetus for further study on the effect of dietary fat on biliary lipid composition comes from a recent report suggesting an increased prevalence of cholesterol gallstones in men on a long-term, high polyunsaturated fat diet (121).

During the overnight fast even nongallstone patients develop bile that is saturated with cholesterol. Also, individuals eating only one or two meals per day are subject to prolonged periods of fasting. Thus, the overnight fast and meal frequency may be significant factors in cholesterol gallstone formation.

6. Disease

There are a number of conditions that have been associated with an increased incidence of cholesterol gallstones including ileal disease, diabetes mellitus, cirrhosis, and vagotomy (122). The few studies of gallstone formation under these circumstances have contributed to our understanding of the pathogenesis of cholesterol gallstones.

7. Drugs

Clofibrate is an effective inhibitor of cholesterol synthesis and is used therapeutically to lower serum cholesterol. Studies in hyperlipidemic patients receiving clofibrate have shown that biliary cholesterol secretion and fecal endogenous cholesterol excretion are increased despite a concomitant decrease in cholesterol synthesis (123). In normal humans, clofibrate increases biliary cholesterol saturation while decreasing bile acid synthesis and pool size (124). Moreover, two studies suggest that the incidence of gallstones doubles in normal and hyperlipidemic patients receiving clofibrate treatment (125,126).

Cholestyramine, an anion exchange resin, binds bile acids in the intestine and promotes their fecal excretion. It has been used, therefore, for treatment of hypercholesterolemia and bile acid induced diarrhea. In animal models there is conflicting evidence regarding the effects of cholestyramine on biliary lipid composition and gallstone formation (127,128). Similarly, in man, studies have shown either no change (129,130) or an increase (131) in biliary cholesterol saturation. As yet this agent has not been reported to cause gallstones in man.

III. CHENODEOXYCHOLIC ACID TREATMENT OF GALLSTONES

There are approximately 16 million Americans with gallstones and each year 6000–8000 people die as a result of them (4). Three hundred and fifty thousand cholecystectomies are performed yearly at an expense of almost a billion dollars. Surgery provides an effective and safe treatment in approximately 90% of these patients and has been considered the treatment of choice.

Due to an increasing knowledge about the natural history of gallstones and their complications, clinical concepts of gallstone disease are changing. Cholesterol gallstones are very common, and just as frequently as not, are asymptomatic when discovered. Many will never become symptomatic or only after many years will cause episodes of biliary colic and cholecystitis as their most common manifestations. Because of the small but significant risk of surgery, its expense, and the lack of urgency in treating most gallstone patients, there is a definite place for the medical therapy of gallstone disease.

A. Clinical Trials

The first reports of successful gallstone dissolution by mixtures of exogenous bile acids appeared in 1937 (132) and 1957 (133). In 1971 Thistle and Schoenfield hypothesized that the high cholesterol saturation of bile in gallstone patients was due to the small bile acid pool with a consequent decrease in bile acid secretion and secondarily a decrease in lecithin secretion. In an attempt to expand the bile acid pool they fed gallstone patients chenodeoxycholic acid and produced unsaturated bile (130). Danzinger *et al.* subsequently treated a group of patients with radiolucent gallstones in gallbladders visualized by oral cholecystography and reported successful dissolution in 4 of 7 patients given chenodeoxycholic acid for 6 months to 2 years (134). At the same time Bell *et al.* treated gallstone patients in England and also reported successful dissolution in patients with radiolucent stones on oral cholecystography (149). In the first controlled study, Thistle and Hofmann reported successful dissolution in 11 of 18 patients with radiolucent gallstones treated with chenodeoxycholic acid (135). In this study cholic acid did not decrease biliary cholesterol saturation (130) and was ineffective at gallstone dissolution (135).

Treatment with chenodeoxycholic acid is most effective in dissolution of radiolucent gallstones. Patients with radiopaque gallstones and gallstones with calcified rims have not responded well to treatment with only 2 of 13 patients undergoing dissolution in one study. It should be noted that even with radiolucency as a selection criteria for treatment the response rate will not be greater than 85–90% since 10–15% of radiolucent gallstones are made of bilirubin pigment and would not be expected to respond. Gallbladders that do not visualize by oral cholecystography and, therefore, presumably contain gallstones, do not regain the ability to concentrate x-ray contrast material with treatment. The fate of the gallstones in these nonvisualizing gallbladders is therefore unknown.

B. Mechanism of Action

Schoenfield *et al.* postulated that the mechanism of chenodeoxycholic acid action was expansion of the bile acid pool by the administered chenodeoxycholic acid. At doses of 1 g/day or greater, chenodeoxycholic acid does indeed expand the bile acid pool while decreasing the relative concentration of biliary cholesterol (136) and reducing the duration of secretion of saturated bile (137). Changes in biliary lipid composition in response to 750 mg/day of chenodeoxycholic acid are shown in Fig. 6. Chenodeoxycholic acid at a dose of 500 mg/day, however, does not expand the bile acid pool, yet decreases biliary cholesterol secretion and saturation (138,139). The secretion of total bile acids and lecithin are not affected. Cholic acid by contrast, at the same dose, increases the bile acid pool, but does not change biliary cholesterol secretion or saturation. The mechanism of bile desaturation thus appears to be complicated and involves factors other than the total bile acid pool size.

Adler *et al.* have discussed several possible mechanisms for the chenodeoxycholic acid induced decrease in biliary cholesterol secretion

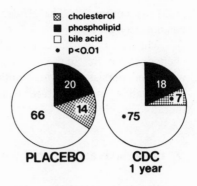

Fig. 6. Biliary lipid composition, expressed as molar percent, in gallstone patients after one year of either chenodeoxycholic acid or placebo treatment. Taken from Schoenfield and Coyne (101).

(138). There is no evidence that chenodeoxycholic acid decreases intestinal cholesterol absorption in either animals (52) or humans (138). At a dose of 750 mg/day chenodeoxycholic acid decreases hepatic HMG-CoA reductase activity in gallstone patients suggesting that there is inhibition of cholesterol synthesis (100). This presumed inhibition of cholesterol synthesis does not appear to result from increased cholesterol absorption or expansion of the cholesterol pool (140,141). However, it is not known whether the decrease in cholesterol secretion is due to chenodeoxycholic acid or a metabolite such as ursodeoxycholic acid (142) or whether the effect of chenodeoxycholic acid is directly on synthesis or secretion of cholesterol or both. Also, it remains to be shown that doses of chenodeoxycholic acid lower than 750 mg/day that decrease cholesterol secretion cause similar inhibition of enzyme activity.

C. Dose

The dose of chenodeoxycholic acid has usually been 750–1000 mg/day. It has recently been shown that 1000 mg/day of chenodeoxycholic acid has no significantly greater effect on biliary lipid composition than 500 mg/day, though more patients develop unsaturated fasting bile with the 1000 mg dose (139). In a one month trial with a different chenodeoxycholic acid preparation, 250 mg, but not 125 mg/day was effective at changing biliary lipid composition (143). Longer treatment at 250 mg/day will be necessary to determine if this dose is as effective as 500 mg/day in altering bile composition. One problem in comparing dosage data from different studies is the possible variation of bioavailability among different preparations.

If one looks at actual gallstone dissolution there appears to be little difference in gallstone dissolution when 750 mg/day (144) and larger doses

(135) are compared. The efficacy of smaller doses in long-term treatment has not been studied.

Marks *et al.* have recently demonstrated that after initiation of chenodeoxycholic acid treatment biliary lipid changes occur within one week and persist for two months following cessation of treatment (143). This finding makes future trials of intermittent chenodeoxycholic acid therapy feasible.

D. Prophylaxis and Retained Stones

Prophylactic therapy with chenodeoxycholic acid may be necessary and should be evaluated in two situations. Racial groups such as the North American Indian, which have a very high incidence of gallstones, might benefit from early and lifelong treatment with chenodeoxycholic acid. Following successful gallstone dissolution and chenodeoxycholic acid withdrawal, biliary cholesterol saturation returns to pretreatment levels (139), and recurrence of gallstones may occur (145). It is not known how frequently patients will have recurrence of their stones, and it may be necessary to continue lifelong treatment following dissolution at least in some patients. Because of this potential need for long term prophylaxis it is important to evaluate low dose and intermittent therapy.

Five to ten percent of patients may have retained stones following cholecystectomy (146). Frequently reoperation is necessary, and the surgery can be very difficult especially if the stones lie within the hepatic bile duct radicles. Oral administration of chenodeoxycholic acid may be successful in dissolving these stones though there is presently no clinical experience with it in this situation.

E. Side Effects

1. Diarrhea

The most frequently observed side effect of chenodeoxycholic acid therapy is diarrhea. Only 50–75% of the currently available preparations is absorbed, with the rest appearing in the stool as chenodeoxycholic acid or lithocholic acid. Elevated concentrations of dihydroxy bile acids in the colon induce a net flux of electrolyte-rich fluid into the bowel (147) probably by activating adenylate cyclase and stimulating cyclic adenosine monophosphate (c-AMP) formation (148). The diarrhea is dose related and at doses of 500–750 mg/day is infrequent or transient.

2. Passage of Stones

Smaller gallstones are more likely than larger ones to pass from the gallbladder into the common duct where they may cause pain, obstruction, infection, and pancreatitis. It is possible, therefore, that patients undergoing chenodeoxycholic acid treatment with progressive decrease in gallstone size are more prone to common duct stones and their complications. Although acute cholecystitis during chenodeoxycholic acid treatment has been reported (149), there is no evidence that any complication of gallstones occurs more frequently in treated than untreated patients. It will be necessary to observe large numbers of gallstone patients in controlled studies to determine the true incidence of complications.

3. Lipid Alterations

There has been concern that chenodeoxycholic acid treatment might increase the total body pool of cholesterol and promote atherosclerosis by increasing cholesterol absorption or decreasing cholesterol conversion to bile acids. Although biliary cholesterol secretion is reduced by chenodeoxycholic acid, cholesterol synthesis is either normal (140) or slightly reduced (141), and absorption (138), excretion (138), and pool size (141,150) are unaltered in man. Serum cholesterol does not increase with treatment though triglycerides may decrease (141,151).

4. Hepatotoxicity

Lithocholic acid, a major fecal bile acid in man, is formed solely in the distal intestine by 7α-dehydroxylation of chenodeoxycholic acid (152). Under normal conditions it has been held that lithocholic acid is little absorbed since it is almost insoluble and is present in very small amounts in bile [less than 4% of total bile acids (95,138)]. Lithocholic acid is a potent hepatotoxin in animals (153), and since a substantial increase in colonic lithocholic acid formation would be expected with chenodeoxycholic acid treatment, concern was expressed very early that such treatment might be associated with hepatotoxicity in man.

In man, doses of chenodeoxycholic acid as high as 4 g/day result in only small increases in biliary lithocholic acid, and lithocholic acid concentrations are almost always less than 8% of the total bile acids. Despite this fact, mild but transient abnormalities in liver function tests (transaminases) occur in up to 1/3 of patients (135,149) and appear to be dose related (139). Liver biopsies have shown only minor and inconsistent microscopic abnormalities not different from those seen in untreated patients. With comparable doses of chenodeoxycholic acid, rhesus monkeys may develop significant morphologic liver changes however, unlike humans, their biliary lithocholic acid reaches levels as high as 14% (150).

There is no conclusive evidence that lithocholic acid is responsible for the abnormal liver function tests seen during chenodeoxycholic acid treatment. If it is responsible it would appear that man is less susceptible to lithocholic acid toxicity than animals. A possible explanation for this is provided by the work of Palmer and Bolt (27) and Cowan *et al.* (28) wherein a significant amount of biliary lithocholic acid (40–75%) was found to be present as the sulfate in humans. There are two important implications of this finding. First, the usual method of bile acid determination, which does not measure sulfated bile acids, would lead to an underestimate of the significance of lithocholic acid absorption. Second, a greater ability to sulfate lithocholic acid might be responsible for the lesser hepatotoxicity of chenodeoxycholic acid in man as compared to other animals if sulfated lithocholic acid is less toxic (152) or more rapidly excreted (28) than lithocholic acid.

Lithocholic acid is not the only endogenously formed metabolite of chenodeoxycholic acid. Ursodeoxycholic acid, present as less than 1% of total bile acids in normal bile, is found in increased amounts in bile of patients receiving chenodeoxycholic acid and has also been shown to be a metabolic product of chenodeoxycholic acid (142). It is not known whether ursodeoxycholic acid is hepatotoxic or, on the contrary, whether is it part of an important detoxification pathway.

Salen *et al.* (154), in a preliminary communication, have reported that rhesus monkeys fed chenodeoxycholic acid and lincomycin (a poorly absorbed antibiotic) have less lithocholic acid in their bile than monkeys fed chenodeoxycholic acid alone and that lincomycin prevents chenodeoxycholic acid induced hepatic lesions. They speculate that increased intestinal lithocholic acid production and absorption is responsible for the hepatic lesions, and lincomycin inhibits bacterial dehydroxylation of chenodeoxycholic acid to lithocholic acid. Thus, a product of the action of bacterial enzymes on chenodeoxycholic acid, perhaps lithocholic acid, may be responsible for chenodeoxycholic acid induced hepatotoxicity.

IV. POTENTIAL AGENTS FOR TREATMENT OF CHOLESTEROL GALLSTONES

There are several agents that decrease biliary cholesterol saturation, but the effectiveness of most of these agents in gallstone dissolution has not been tested. In man, phenobarbital decreases biliary cholesterol saturation and causes disappearance of cholesterol crystals. However, during one year of treatment, fasting gallbladder bile did not become unsaturated in most patients and gallstone dissolution did not occur (144). Although large doses of lecithin have been proposed as a possible agent for gallstone dissolution (45), controversy exists about the ability of lecithin to alter biliary lipid composition in man (44,45), and there are no reports of its efficacy in gallstone dissolution. Glycerophosphate, a precursor of phospholipids, decreases biliary cholesterol saturation in gallstone patients, but treatment with 6 g/day for 6 days did not produce unsaturated bile (155). Thyroid hormone has been reported to increase the synthesis of both primary bile acids and increase the ratio of chenodeoxycholic acid to cholic acid in bile (156,157), but its effect on biliary lipid composition is not known.

Several dietary manipulations have potential therapeutic value and need further study. Bran increases the percentage of chenodeoxycholic acid in bile (158) and may improve biliary cholesterol solubility. Tricaprylin, a medium-chain triglyceride, increases the bile acid recycling frequency, decreases the bile acid pool size, and decreases biliary cholesterol saturation in rhesus monkeys (23).

Although many of these therapeutic agents and dietary manipulations reduce cholesterol saturation without desaturating bile it is possible that a combination of agents will be effective in desaturating bile. In addition they may also be effective for prophylaxis.

REFERENCES

1. R. H. Palmer and Z. Hruban, *J. Clin. Invest.* **45**, 1255 (1966).
2. F. G. Zaki, J. B. Carey, F. W. Hoffbauer, and C. Nwokolo, *J. Lab. Clin. Med.* **69**, 737 (1967).
3. H. Miyake and C. G. Johnston, *Digestion* **1**, 210 (1968).
4. F. J. Ingelfinger, *Gastroenterology* **55**, 102 (1968).
5. J. M. Dietschy and J. D. Wilson, *N. Engl. J. Med.* **282**, 1128 (1970).
6. H. Danielsson and T. T. Tchen, *in* "Metabolic Pathways" (D. M. Greenberg, ed.), Vol. 2, p. 117, Academic Press, New York (1968).
7. M. D. Siperstein and V. M. Fagan, *J. Biol. Chem.* **241**, 602 (1966).
8. H. S. Sodhi, R. C. Orchard, N. D. Agnish, P. U. Varughese, and B. J. Kudchodkar, *Atherosclerosis* **17**, 197 (1973).
9. S. M. Grundy, A. F. Hofmann, J. Davignon, and E. H. Ahrens, Jr., *J. Clin. Invest.* **45**, 1018 (1966).
10. S. M. Grundy and E. H. Ahrens, Jr., *J. Clin. Invest.* **45**, 1503 (1966).
11. J. D. Wilson and C. A. Lindsey, Jr., *J. Clin. Invest.* **44**, 1805 (1965).
12. W. E. Connor and D. S. Lin, *J. Clin. Invest.* **53**, 1062 (1974).

13. E. Quintão, S. M. Grundy, and E. H. Ahrens, Jr., *J. Lipid Res.* **12**, 233 (1971).
14. R. P. Cook, A. Kliman, and L. F. Fieser, *Arch. Biochem.* **52**, 439 (1954).
15. P. J. Nestel, H. M. Whyte, and D. S. Goodman, *J. Clin. Invest.* **48**, 982 (1969).
16. E. P. M. Bhattathiry and M. D. Siperstein, *J. Clin. Invest.* **42**, 1613 (1963).
17. T. A. Miettinen, *Circulation* **54**, 842 (1971).
18. W. E. Connor, D. T. Witiak, D. B. Stone, and M. L. Armstrong, *J. Clin. Invest.* **48**, 1363 (1969).
19. S. M. Grundy and E. H. Ahrens, Jr., *J. Clin. Invest.* **49**, 1135 (1970).
20. H. S. Sodhi, B. J. Kudchodkar, L. Horlick, and C. H. Weder, *Metabolism* **20**, 348 (1971).
21. D. H. Gregory, Z. R. Vlahcevic, P. Schatzki, and L. Swell, *J. Clin. Invest.* **55**, 105 (1975).
22. L. Den Besten, W. E. Connor, and S. Bell, *Surgery* **73**, 266 (1973).
23. R. N. Redinger, A. H. Herman, and D. M. Small, *Gastroenterology* **64**, 610 (1973).
24. C. K. McSherry, K. P. Morrissey, N. B. Javitt, and F. Glenn, *Ann. Surg.* **178**, 669 (1973).
25. P. H. Schreibman, D. Pertsemlidis, G. C. K. Liu, and E. H. Ahrens, Jr., *J. Clin. Invest.* **53**, 72a (1974) (Abstr.).
26. E. H. Mosbach, *Arch. Intern. Med.* **130**, 478 (1972).
27. R. H. Palmer and M. G. Bolt, *J. Lipid Res.* **12**, 671 (1971).
28. A. E. Cowen, M. G. Korman, and A. F. Hofmann, *Gastroenterology* **67**, 8 (1974) (Abstr.).
29. E. Krag and S. F. Phillips, *J. Clin. Invest.* **53**, 1686 (1974).
30. G. W. Hepner, A. F. Hofmann, and P. J. Thomas, *J. Clin. Invest.* **51**, 1898 (1972).
31. B. Borgström, A. Dahlquist, G. Lundh, and J. Sjövall, *J. Clin. Invest.* **36**, 1521 (1957).
32. H. Danielsson, P. Eneroth, K. Hellström, S. Lindstedt, and J. Sjövall, *J. Biol. Chem.* **238**, 2299 (1963).
33. S. M. Grundy, E. H. Ahrens, Jr., and G. Salen, *J. Lab. Clin. Med.* **78**, 94 (1971).
34. S. Balasubramaniam, K. A. Mitropoulos, and N. B. Myant, *Biochim. Biophys. Acta* **326**, 428 (1973).
35. J. H. Gans and M. R. Cater, *Am. J. Physiol.* **217**, 1018 (1969).
36. S. M. Grundy, *J. Clin. Invest.* **55**, 269 (1975).
37. S. Nilsson, *Acta Chir. Scand.* **405**, Suppl. 1 (1970).
38. J. A. Balint, D. A. Beeler, D. H. Treble, and H. L. Spitzer, *J. Lipid Res.* **8**, 486 (1967).
39. P. Björnstad and J. Bremer, *J. Lipid Res.* **7**, 38 (1966).
40. J. A. Balint, D. A. Beeler, E. C. Kyriakides, and D. H. Treble, *J. Lab. Clin. Med.* **77**, 122 (1971).
41. B. Borgström, *Acta Chem. Scand.* **11**, 749 (1957).
42. B. Arnesjö, R. Nilsson, J. Barrowman, and B. Borgström, *Scand. J. Gastroenterol.* **4**, 653 (1969).
43. D. R. Saunders, *Gastroenterology* **59**, 848 (1970).
44. J. L. Thistle and L. J. Schoenfield, *Clin. Res.* **16**, 450 (1968).
45. R. K. Tompkins, G. B. Burke, R. M. Zollinger, and D. G. Cornwell, *Ann. Surg.* **172**, 936 (1970).
46. C. K. McSherry, K. P. Morrissey, N. B. Javitt, and F. Glenn, *Ann. Surg.* **178**, 669 (1973).
47. L. M. van Golde and L. L. M. van Deenen, *Biochim. Biophys. Acta* (Amst.) **125**, 496 (1966).
48. I. M. van Golde, W. A. Pieterson, and L. L. M. van Deenen, *Biochim. Biophys. Acta* (Amst.) **152**, 84 (1968).

49. E. Cahlin, J. Jönsson, S. Nilsson, and T. Scherstén, *Scand. J. Clin. Lab. Invest.* **29,** 109 (1972).
50. M. D. Siperstein, I. L. Chaikoff, and W. O. Reinhardt, *J. Biol. Chem.* **198,** 111 (1952).
51. C. R. Treadwell and G. V. Vahouny, *in* "Handbook of Physiology" (C. F. Code, ed.), Vol. 3, p. 1407, American Physiological Society, Washington (1968).
52. R. F. Raicht, B. I. Cohen, and E. H. Mosbach, *Gastroenterology* **67,** 1155 (1974).
53. J. D. Wilson, *Arch. Intern. Med.* **130,** 493 (1972).
54. J. M. Dietschy and M. D. Siperstein, *J. Clin. Invest.* **44,** 1311 (1965).
55. J. M. Dietschy, *J. Clin. Invest.* **47,** 286 (1968).
56. J. W. Huff, J. L. Gilfillan, and J. M. Hunt, *Proc. Soc. Exp. Biol. Med.* **114,** 352 (1963).
57. W. T. Beher, G. D. Baker, W. L. Anthony, and M. E. Better, *Henry Ford Hosp. Med. Bull.* **9,** 201 (1961).
58. D. C. Young and K. C. Hanson, *J. Lipid Res.* **13,** 244 (1972).
59. T. Scherstén, *Digestion* **9,** 540 (1973).
60. R. H. Dowling, E. Mack, and D. M. Small, *J. Clin. Invest.* **50,** 1917 (1971).
61. D. M. Small, *Mol. Cryst. Liq. Cryst.* **8,** 209 (1969).
62. R. H. Dowling, E. Mack, J. Picott, J. Berger, and D. M. Small, *J. Lab. Clin. Med.* **72,** 169 (1968).
63. C. D. Klaassen, *Eur. J. Pharmacol.* **23,** 270 (1973).
64. S. Erlinger, D. Dhumeaux, and J. P. Benhamous, *Nature* **223,** 1276 (1969).
65. M. C. Carey and D. M. Small, *Arch. Intern. Med.* **130,** 506 (1972).
66. W. H. Admirand and D. M. Small, *J. Clin. Invest.* **47,** 1043 (1968).
67. R. T. Holzbach, M. Marsh, M. Olszewski, and K. Holan, *J. Clin. Invest.* **52,** 1467 (1973).
68. R. T. Holzbach and C. Y. C. Pak, *Am. J. Med.* **56,** 141 (1974).
69. M. F. Olszewski and R. T. Holzbach, *Nature* **242,** 336 (1973).
70. A. L. Metzger, S. Heymsfield, and S. M. Grundy, *Gastroenterology* **62,** 499 (1972).
71. P. J. Thomas and A. F. Hofmann, *Gastroenterology* **65,** 698 (1973).
72. H. Bogren, *Acta Radiol.* (*Diagn.*) *Suppl.* **226,** 1 (1964).
73. I. S. Russel, M. B. Wheeler, and R. Freake, *Br. J. Surg.* **55,** 161 (1968).
74. N. A. Womack, R. Zeppa, and G. L. Irvin, *Ann. Surg.* **157,** 670 (1963).
75. F. Nakayama, *J. Lab. Clin. Med.* **77,** 366 (1971).
76. K. Juniper, Jr. and E: N. Burson, Jr., *Gastroenterology* **32,** 175 (1957).
77. R. T. Holzbach, M. F. Olszewski, and M. Marsh, *J. Clin. Invest.* **52,** 41 (1973) (Abstract.)
78. W. I. Higuchi, S. Prakongpan, V. Surpuriya, F. Young, *Science* **178,** 633 (1972).
79. T. Osuga, K. Mitamura, S. Miyagawa, N. Sato, S. Kintaka, and O. W. Portman, *Lab. Invest.* **31,** 696 (1975).
80. D. J. Sutor and S. E. Wooley, *Gut* **15,** 130 (1974).
81. R. N. Redinger and D. M. Small, *Arch. Intern. Med.* **130,** 618 (1972).
82. C. C. Bell, Jr., Z. R. Vlahcevic, J. Prazich, and L. Swell, *Surg. Gynecol. Obstet.* **136,** 961 (1973).
83. D. M. Small and S. Rapo, *N. Engl. J. Med.* **283,** 53 (1970).
84. S. M. Grundy, W. C. Duane, R. D. Adler, J. M. Aron, and A. L. Metzger, *Metabolism* **23,** 67 (1974).
85. T. C. Northfield and A. F. Hofmann, *Gut* **16,** 1 (1975).
86. Z. R. Vlahcevic, C. C. Bell, Jr., J. Buhac, J. T. Farrar, and L. Swell, *Gastroenterology* **59,** 165 (1970).
87. E. W. Pomare and K. W. Heaton, *Gut* **14,** 885 (1973).
88. A. L. Baker, M. M. Kaplan, R. A. Norton, and J. F. Patterson, *Am. J. Dig. Dis.* **19,** 109 (1974).

89. J. F. Woodbury and F. Kern, Jr., *J. Clin. Invest.* **50**, 2531 (1971).
90. P. Abaurre, S. G. Gordon, J. G. Mann, and F. Kern, Jr., *Gastroenterology* **57**, 679 (1969).
91. R. H. Dowling, G. D. Bell, and J. White, *Gut* **13**, 415 (1972).
92. L. J. Schoenfield, J. Sjövall, and K. Sjövall, *J. Lab. Clin. Med.* **68**, 186 (1966).
93. A. Mallory, F. Kern, Jr., J. Smith, and D. Savage, *Gastroenterology* **64**, 26 (1973).
94. T. S. Low-Beer and E. W. Pomare, *Br. Med. J.* **2**, 338 (1973).
95. R. G. Danzinger, A. F. Hofmann, J. L. Thistle, and L. J. Schoenfield, *J. Clin. Invest.* **52**, 2809 (1973).
96. W. J. Banfield and W. H. Admirand, *Clin. Res.* **23**, 245a (1975) (Abstr.).
97. H. R. Almond, Z. R. Vlahcevic, C. C. Bell, D. H. Gregory, and L. Swell, *N. Engl. J. Med.* **289**, 1213 (1973).
98. S. M. Grundy, A. L. Metzger, and R. D. Adler, *J. Clin. Invest.* **51**, 3026 (1972).
99. G. Nicolau, S. Shefer, G. Salen, and E. H. Mosbach, *J. Lipid Res.* **15**, 146 (1974).
100. G. G. Bonorris, M. J. Coyne, L. I. Goldstein, and L. J. Schoenfield, *Gastroenterology* **67**, 780 (1974) (Abstr.).
101. L. J. Schoenfield, L. I. Goldstein, and L. Kaplan, Diseases of the gallbladder and biliary system, Medcom Slide Series, Medcom, Inc., New York (1974).
102. G. Nicolau, S. Shefer, G. Salen, and E. H. Mosbach, *J. Lipid Res.* **15**, 94 (1974).
103. L. Swell, C. C. Bell, and Z. R. Vlahcevic, *Gastroenterology* **61**, 716 (1971).
104. E. A. Shaffer, J. W. Braasch, and D. M. Small, *N. Engl. J. Med.* **287**, 1317 (1972).
105. F. Simmons, A. P. J. Ross, and I. A. D. Bouchier, *Gastroenterology* **63**, 466 (1972).
106. R. D. Adler, A. L. Metzger, and S. M. Grundy, *Gastroenterology* **66**, 1212 (1974).
107. J. R. Malagelada, V. L. W. Go, W. H. J. Summerskill, and W. Gamble, *Am. J. Dig. Dis.* **18**, 455 (1973).
108. E. Thureborn, *Gastroenterology* **50**, 775 (1966).
109. M. M. Gerdes and E. A. Boyden, *Surg. Gynecol. Obstet.* **66**, 145 (1938).
110. J. D. Ostrow, *J. Lab. Clin. Med.* **74**, 482 (1969).
111. D. H. Neiderhiser, W. A. Morningstar, and H. P. Roth, *J. Lab. Clin. Med.* **82**, 891 (1973).
112. J. D. Ostrow, *J. Lab. Clin. Med.* **78**, 255 (1971).
113. D. A. Neiderhiser, F. M. Pineda, L. J. Hejduk, and H. P. Roth, *J. Lab. Clin. Med.* **71**, 985 (1971).
114. C. E. Jackson and B. C. Gay, *Surgery* **46**, 853 (1959).
115. R. G. Danzinger, H. Gordon, L. J. Schoenfield, and J. L. Thistle, *Mayo Clin. Proc.* **47**, 762 (1972).
116. R. E. Sampliner, P. H. Bennett, L. J. Comess, F. A. Rose, and T. A. Burch, *N. Engl. J. Med.* **283**, 1358 (1970).
117. M. M. Fischer and I. M. Yousef, *Can. Med. Assoc. J.* **109**, 190 (1973).
118. Boston Collaborative Drug Surveillance Program. Oral Contraceptives and Venous Thromboembolic Disease, Surgically Confirmed Gallbladder Disease and Breast Tumors, *Lancet* **1**, 1399 (1973).
119. Boston Collaborative Drug Surveillance Program. Surgically Confirmed Gallbladder Disease, Venous Thromboembolisms, and Breast Tumors in Relation to Postmenopausal Estrogen Therapy, *N. Engl. J. Med.* **290**, 15 (1974).
120. G. D. Friedman, W. B. Kannel, and T. R. Dawber, *J. Chronic. Dis.* **19**, 273 (1966).
121. R. A. L. Sturdevant, M. L. Pearce, and S. Dayton, *N. Engl. J. Med.* **288**, 24 (1973).
122. M. D. Kaye and F. Kern, Jr., *Lancet* **1**, 1228 (1971).
123. S. M. Grundy, E. H. Ahrens, Jr., G. Salen, P. H. Schreibman, and P. J. Nestel, *J. Lipid Res.* **13**, 531 (1972).

124. D. Pertsemlidis, A. Panveliwalla, and E. H. Ahrens, Jr., *Gastroenterology* **66**, 565 (1974).
125. The Coronary Drug Project, *J.A.M.A.* **231**, 360 (1975).
126. J. Cooper, H. Geizerova, and M. F. Oliver, *Lancet* **1**, 1083 (1975).
127. R. Bergman, and W. Van der Linden, *Gastroenterology* **53**, 418 (1967).
128. L. J. Schoenfield and J. Sjövall, *Am. J. Physiol.* **211**, 1069 (1966).
129. W. Van der Linden and F. Nakayama, *Acta. Chir. Scand.* **135**, 433 (1969).
130. J. L. Thistle and L. J. Schoenfield, *Gastroenterology* **61**, 488 (1971).
131. H. Dam, I. Prange, M. K. Jensen, H. E. Kallehauge, and H. J. Fenger, *Z. Ernährungswiss.* **10**, 188 (1971).
132. A. G. Rewbridge, *Surgery* **1**, 395 (1937).
133. W. H. Cole and W. H. Harridge, *J. Am. Med. Assoc.* **164**, 238 (1957).
134. R. G. Danzinger, A. F. Hofmann, L. J. Schoenfield, and J. L. Thistle, *N. Engl. J. Med.* **286**, 1 (1972).
135. J. L. Thistle and A. F. Hofmann, *N. Engl. J. Med.* **289**, 655 (1973).
136. L. J. Schoenfield, R. G. Danzinger, A. F. Hofmann, J. L. Thistle, *In* "The Liver. Quantitative Aspects of Structure and Function," p. 174, Karger, Basel (1973).
137. T. C. Northfield, N. F. LaRusso, A. F. Hofmann, and J. L. Thistle, *Gut* **16**, 12 (1975).
138. R. D. Adler, L. J. Bennion, W. C. Duane, and S. M. Grundy, *Gastroenterology* **68**, 326 (1975).
139. H. Y. I. Mok, G. D. Bell, and R. H. Dowling, *Lancet* **2**, 253 (1974).
140. N. E. Hoffman, A. F. Hofmann, and J. L. Thistle, *Mayo Clin. Proc.* **49**, 236 (1974).
141. L. Pedersen, T. Arnfred, and E. H. Thaysen, *Scand. J. Gastroenterol.* **9**, 787 (1974).
142. G. Salen, G. S. Tint, B. Eliav, N. Deering, and E. H. Mosbach, *J. Clin. Invest.* **53**, 612 (1974).
143. J. Marks, G. Bonorris, A. Chung, M. J. Coyne, L. I. Goldstein, R. Okun, and L. J. Schoenfield, *Gastroenterology* **68**, 89 (1975) (Abstr.).
144. M. J. Coyne, G. G. Bonorris, A. Chung, L. I. Goldstein, D. Lahana, and L. J. Schoenfield, *N. Engl. J. Med.* **292**, 604 (1975).
145. J. L. Thistle, P. Y. S. Yu, A. F. Hofmann, and B. J. Ott, *Gastroenterology* **66**, 789 (1974).
146. R. B. Smith and F. A. Conklin, *Surg. Gynecol. Obstet.* **116**, 731 (1963).
147. H. S. Mekhjian, S. F. Phillips, and A. F. Hofmann, *J. Clin. Invest.* **50**. 1569 (1971).
148. D. R. Conley, M. J. Coyne, A. Chung, G. G. Bonorris, and L. J. Schoenfield, *Gastroenterology* **68**, 20 (1975) (Abstr.).
149. G. D. Bell, B. Whitney, and R. H. Dowling, *Lancet* **2**, 1213 (1973).
150. K. H. Webster, M. C. Lancaster, A. F. Hofmann, D. F. Wease, and A. H. Baggenstoss, *Mayo Clin. Proc.* **50**, 134 (1975).
151. G. D. Bell, B. Lewis, A. Petrie, and R. H. Dowling, *Br. Med. J.* **3**, 520 (1973).
152. R. H. Palmer, *Arch. Intern. Med.* **130**, 606 (1972).
153. R. D. Hunt, G. A. Leveille, and H. E. Sauberlich, *Proc. Soc. Exp. Biol. Med.* **115**, 277 (1964).
154. G. Salen, H. Dyrszka, T. Chen, W. H. Saltzman, and E. H. Mosbach, *Lancet* **1**, 1082 (1975).
155. W. G. Linscheer and K. L. Raheja, *Lancet* **2**, 551 (1974).
156. W. W. Faloon, *Am. J. Dig. Dis.* **19**, 81 (1974).
157. K. Hellström and S. Lindstedt, *J. Lab. Clin. Med.* **63**, 666 (1964).
158. E. W. Pomare, K. W. Heaton, *Br. Med. J.* **4**:262 (1973).

Chapter 6

THE METABOLISM OF STEROLS AND BILE ACIDS IN CEREBROTENDINOUS XANTHOMATOSIS*

G. Salen

College of Medicine and Dentistry of New Jersey
New Jersey Medical School, Newark, New Jersey;
and East Orange V.A. Hospital, East Orange, New Jersey

and

E. H. Mosbach

Department of Lipid Research
The Public Health Research Institute of the City of New York, Inc.

I. INTRODUCTION*

Xanthomas are defined as localized collections of lipids that deposit in certain areas of the body. They are most frequently found in Achilles and dorsal tendons of hands (tendon) and skin (tuberous). Their presence signifies a disorder of lipid metabolism and presumably indicates that plasma cholesterol or triglyceride levels are elevated. In the hypercholesterolemic states, plasma cholesterol levels generally exceed 300 mg/100 ml (1,2,3). Although xanthomas in themselves do not produce serious symptoms except for cosmetic deformity and perhaps difficulty in mobility caused by the infiltration of the xanthomatous deposit in the tendon, their presence suggests that excessive amounts of cholesterol have accumulated in other tissues. The development of ischemic symptoms caused by atherosclerotic lesions that progressively stenose vessels is a well-known complication. The

* Supported in part by Research Grants No. AM-05222, HE-10894, HL-17818 from the U.S.P.H.S and BMS 75-01168 from the N.S.F.

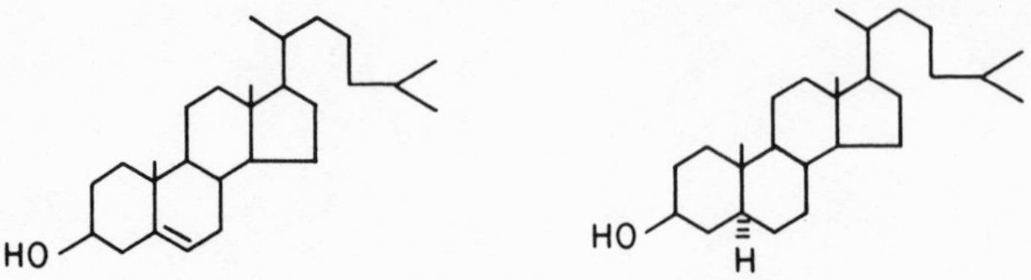

CHOLESTEROL   CHOLESTANOL

Fig. 1. Structures of cholesterol and cholestanol (5α-cholestan-3β-ol).

cholesterol apparently is deposited from the plasma lipoproteins; the force promoting the depositions is the level of plasma cholesterol (4). A widely held theory is that the level of plasma cholesterol is of prime importance in promoting the flux of plasma cholesterol into the tissue. A corollary of this point of view is that lowering of plasma cholesterol levels will be beneficial by decreasing the rate at which cholesterol is deposited in the tissue. Despite the abundance of evidence in favor of this theory, a number of subjects have been described recently who apparently contradict and challenge the idea that the absolute level of plasma cholesterol controls tissue cholesterol deposits, in particular xanthoma formation. Two syndromes have been described where tendon and tuberous xanthomas develop in the presence of low plasma cholesterol levels but with the circulation of increased amounts of other unusual sterols. In one syndrome, the abnormal sterol is β-sitosterol, a naturally occurring plant sterol that differs from cholesterol by the addition of an ethyl group at carbon-24 of the side chain. The case of two sisters with β-sitosterolemia, tuberous xanthomas of the elbows, and xanthomas of the hands, patellar, and Achilles tendons has been described by Bhattacharyya and Connor (5). Increased plasma concentrations of β-sitosterol (18 and 27 mg/100 ml versus a normal level of < 1 mg/100 ml) were present in both sisters, and the β-sitosterol was thought to arise by increased intestinal absorption from the diet. The second disorder of xanthomas developing in the presence of low plasma cholesterol levels is designated cerebrotendinous xanthomatosis (CTX). This rare condition was first described by Van Bogaert in 1937 and the clinical signs consist of progressive neurologic dysfunction, cataracts, premature atherosclerosis, mild pulmonary insufficiency, and adrenal hypofunction (6,7). Plasma levels of cholestanol (the 5α-dihydro derivative of cholesterol) (Fig. 1) are elevated. This report summarizes our clinical and biochemical experience with five patients having the CTX syndrome.

II. CLINICAL MANIFESTATIONS

In an earlier report, the frequency of clinical manifestation of CTX found in six individuals from three families was presented (Table I) (7). Achilles tendon xanthomas were present in all six subjects and represent the commonest clinical manifestation. The xanthomas appeared before other clinical signs and were histologically indistinguishable from those seen in subjects with familial hypercholesterolemia (Type IIa, hyperbetalipoproteinemia). In addition, tuberous xanthomas developed in scar tissue of one subject (J.C.) and it was noted that the tendon xanthoma of one patient (E.D.E.) enlarged after biopsy. Neurological dysfunction was present in four of the six subjects and included dementia, cerebellar ataxia, and corticospinal tract paresis. Two individuals died, and post mortem examination revealed large xanthomatous deposits in the cerebral cortex and cerebellum. Pulmonary dysfunction was detected in four of six subjects and included lung infiltrates found by chest X-ray and abnormal pulmonary function tests; yet pulmonary symptoms were mild. At post mortem, small lipid deposits were present in the lung tissue of M.D. which suggested that xanthomatous lesions may have also developed in the lungs and were probably responsible for the diminished pulmonary function. Coronary atherosclerosis was evidenced in three subjects; two individuals, V.N. and M.D., died of acute myocardial infarctions and, at post mortem, extensive atheromatous lesions were found in the major coronary vessels. In another subject (E.D.S.), repeated episodes of supraventricular tachycardia accompanied by EKG evidence of myocardial ischemia were documented. Cataracts were found only in two subjects (V.N. and G.C.B.). Acute cholecystitis associated with a gallstone was proven at surgery in E.D.S., while cholesterolosis of the gallbladder was found at post mortem in M.D. In the "D" family, 16 additional family members were examined carefully, including the mother, three brothers, and 12 children (six from symptomatic parents and six from asymptomatic parents), and showed neither clinical nor biochemical abnormalities. These observations suggest that the clinical syndrome may evolve slowly in an irregular fashion and thus the diagnosis of the disorder cannot be excluded because characteristic clinical (neurological) manifestations are absent.

Since the original description of this condition by Van Bogaert et al., 17 additional well-documented cases exhibiting the CTX syndrome have been published in the literature plus cases of three individuals who exhibit clinical pictures similar to the CTX syndrome (7–21). However, the authors have personal knowledge of six other individuals with this condition.

TABLE I. Clinical Features[a]

| Patients | Sex | Age | Neurological abnormalities | | | Cataracts | Xanthomatosis | | Pulmonary insufficiency | Endocrine abnormalities | | Clinical atherosclerosis |
			Dementia	Cerebellar ataxia	Paresis corticospinal tract dysfunction		Tendons	Skin		Hypoadrenalism	Hypothroidism	
M.D.[b]	M	36	++	++++	++++	0	++++	0	++	++++	++++	++++
E.D.S.	F	46	0	0	++	0	++++	0	0	0	0	++
E.D.E.	F	44	0	0	+++	0	++++	0	0	0	0	0
J.C.	M	32	0	0	0	0	++++	++	+	0	0	0
G.C.B.	F	34	0	0	0	+++	++++	0	+	0	0	0
V.N.[b]	M	48	++++	++++	++++	++++	++++	0	++	0	0	++++

[a] 0 to ++++ indicates increasing severity.
[b] Deceased.

Therefore, it seems that the CTX syndrome is more common than originally suspected.

III. BIOCHEMICAL FINDINGS

It was appreciated for many years that the tendon and brain xanthomas which characterize CTX develop in the presence of low plasma cholesterol levels. However, it was not until 1969, that Menkes and colleagues observed that increased amounts of cholestanol (5α-cholestan-3β-ol) (Fig. 1) accumulated in the brain tissue of a CTX subject (22). This finding provided a unique biochemical marker for patients with this condition. Plasma and erythrocytes from six symptomatic (and one asymptomatic) CTX subjects, 15 asymptomatic close family members, and 10 normolipidemic control subjects were analyzed for cholesterol and cholestanol (7). The results are presented in Table II. Average plasma cholesterol levels in the group of CTX individuals were significantly lower than in the normal subjects (144 mg/100 ml versus 225 ± 25 mg/100 ml), but were not different from the unaffected relatives. The cholesterol concentrations in the packed erythrocytes from the CTX patients were similar to the levels in the normal controls and the asymptomatic relatives. Plasma cholestanol levels were higher in all the CTX individuals that were tested as compared with the control subjects and were up to 20 times higher than those of the asymptomatic relatives. Similarly, erythrocyte cholestanol concentrations were higher in the CTX subjects than the normal individuals or the asymptomatic members of the "D" family. One member of this family, subject B.D., who is asymptomatic, showed elevated levels of plasma and erythrocyte cholestanol, and we believe that he will eventually develop the disease.

IV. INHERITANCE

The pedigree of the "D" family is presented in Fig. 2 (7). Three of six siblings were symptomatic (Table I) while four of six had increased plasma and erythrocyte cholestanol levels (Table II). The father of the sibship died of an acute myocardial infarction several years ago, but neurological impairment or tendon xanthomas were never evident. The surviving parent and two children from the sibship did not exhibit symptoms nor did they have abnormal cholestanol concentrations. Therefore, if plasma cholestanol levels are related to the development of the clinical syndrome, then the absence of

TABLE II. Plasma and Erythrocyte Sterol Concentrations

Subject	Plasma			Erythrocyte		
	Cholesterol, mg/100 ml	Cholestanol, mg/100 ml	Cholestanol, %	Cholesterol, mg/100 ml[a]	Cholestanol, mg/100 ml[a]	Cholestanol, %
10 Normal subjects (±SD)	225 ± 25	0.9 ± 0.2	0.4	120 ± 10	0.24 ± 0.04	0.2
M.D.	150	—	—	—	—	—
E.D.S.	127	2.2	1.8	111	1.4	1.3
E.D.E.	117	2.2	2.1	113	1.6	1.4
J.C.	129	1.3	1.0	113	2.4	2.1
G.C.B.	144	1.4	1.0	—	—	—
V.N.	196	3.9	2.0	—	—	—
B.D.[b]	141	1.6	1.1	—	—	—
15 Asymptomatic relatives family D (±SD)	126 ± 23	0.24 ± 0.20	0.19	117	0.33	0.27

[a] Packed cells.
[b] Asymptomatic, family "D".

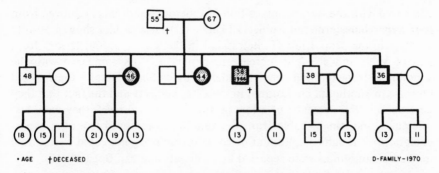

• AGE † DECEASED D-FAMILY-1970

Fig. 2. Pedigree of family in which three members had characteristic manifestations of cerebrotendinous xanthomatosis (cross-hatched). Elevated plasma cholestanol concentrations (heavy lines) were also found in the three affected siblings plus an additional asymptomatic brother. The father of the affected family died of an acute myocardial infarction but did not manifest any of the typical stigmata of the disease.

this biochemical finding in all offspring from the affected sibship, coupled with the lack of clinical findings in both parents, suggests that the disease was inherited as an autosomal recessive trait.

V. TISSUE CONCENTRATIONS OF CHOLESTEROL AND CHOLESTANOL IN CTX

A. Brain Tissue

Table III compares values for Patient V.N. (CTX) with those of a normal man who died unexpectedly (7). More than 200 times more cholestanol was deposited in the frontal lobes and about 900 times as much in the cerebellum as was found in the normal brain; 3–4 times as much cholesterol was also deposited in the CTX brain tissue as compared with brain tissue from the normal man. These studies thus confirm the earlier observations of Menkes et al. (22) who first discovered the increased storage of cholestanol in the brain tissue of an afflicted CTX man; as well as the subsequent reports of Derby et al. (16) and of Phillipart and van Bogaert (15), who also found increased cholestanol in brain tissue of the CTX subjects they studied.

B. Sterol Composition in Xanthomas

The relative amounts of cholesterol and cholestanol in xanthomas removed from the lung, skin, and Achilles tendon of two CTX subjects were compared with the sterol composition of tuberous xanthomas removed from four hypercholesterolemic subjects (Table III). The results show a 50-fold increase in the proportion of cholestanol in the xanthomatous lesions from patient V.N. and a 3-fold increase in patient J.C. as compared with the xanthomas from the hypercholesterolemic individuals. Despite the marked increase in xanthoma cholestanol in the CTX subjects and the fact that they developed in the presence of a low plasma cholesterol level, they were histologically indistinguishable from the xanthomas removed from the hypercholesterolemic subjects. Similar observations showing increased cholestanol in tendon xanthomas were reported by Phillipart and van Bogaert (15).

C. Sterol Composition in Tissues Other Than Brain

The concentrations of cholesterol and cholestanol in 11 histologically normal tissues from a man who died suddenly and unexpectedly were compared with similar measurements from 11 corresponding tissues obtained from a patient (V.N.) with CTX. The results are presented in Table IV (7). Cholestanol was present in all normal tissues examined except lymph nodes; the highest proportions were found in cardiac muscle sterols (0.7%), adrenal gland sterols (0.5%), and kidney sterols (0.4%); the other normal tissue sterols contained about 0.2% cholestanol. In the nonxanthomatous tissues from the CTX patient, cholestanol averaged about 2% of the tissue sterols, with the highest amounts in the atheromatous plaque (2.8%), adipose tissue

TABLE III. Sterol Composition of Brain and Xanthomas

Tissue	Concentration of sterol per wet weight, mg/g	Cholesterol, %	Cholestanol, %
Frontal lobe			
Normal value	9.5	99.7	0.3
Patient V.N.	33.7	80.0	20.0
Cerebellum			
Normal value	8.7	99.9	0.1
Patient V.N.	25.9	66.0	34.0
Xanthomas			
Four patients with hypercholesterolemia (tuberous)	17.6	99.7	0.3
Patient V.N. (tendon)	22.3	88.9	11.1
Patient J.C. (tuberous)	1.9	92.7	7.3
Pulmonary xanthomas			
Patient V.N.	34.2	92.8	7.8

TABLE IV. Sterol Composition of Tissues Other than Brain

Tissue	Normal subject			Patient V.N.		
	Concentration of sterol per wet weight, mg/g	Cholesterol, %	Cholestanol, %	Concentration of sterol per wet weight, mg/g	Cholesterol, %	Cholestanol, %
Lung	2.02	99.8	0.2	1.45	98.1	1.9
Liver	1.20	99.8	0.2	2.10	98.0	2.0
Aorta	—	—	—	3.08	98.2	1.8
Spleen	0.93	99.8	0.2	3.29	98.0	2.0
Kidney	2.44	99.6	0.4	1.50	97.6	2.4
Adipose tissue	1.07	99.8	0.2	23.18	98.0	2.0
Adrenal gland	21.35	99.5	0.5	3.66	98.0	2.0
Testicle	—	—	—	1.18	97.3	2.7
Rectus muscle	0.56	99.8	0.2	57.00	97.2	2.8
Atheromatous plaque	11.21	99.8	0.2	1.70	98.9	1.1
Cardiac muscle	1.02	99.3	0.7	1.81	98.9	1.1
Lymph node	0.97	100	0	—	—	—
Skin	1.10	99.8	0.2			

(2.4%), and rectus muscle (2.7%). Although these tissues were histologically indistinguishable from normal, they contained 3—100 times more cholestanol. Thus, the elevated cholestanol levels in this patient were not restricted to the brain and xanthomatous lesions but actually involved every body tissue, and indicate that the cholestanolosis is a generalized manifestation of CTX. Furthermore, every tissue except plasma and erythrocytes contained significantly more cholesterol than corresponding normal tissues, which suggests an associated defect in cholesterol metabolism (see below). It is important to emphasize that despite the striking elevations in tissue cholestanol concentrations, the increase in tissue cholesterol content was quantitatively more important because cholesterol was always the predominant tissue sterol. Interestingly, the deposition of cholesterol occurred in the presence of low plasma concentrations. This finding, as we have pointed out before, is in conflict with current views that xanthomas in hypercholesterolemic diseases are formed from the imbibition of plasma cholesterol in areas under stress; the rate of development being related to the level of plasma cholesterol. Perhaps the higher levels of plasma cholestanol and circulating bile alcohols (see below) affect the stability of cholesterol and cholestanol in the circulating lipoprotein complex and favor the deposition of the sterols into the tissues.

D. Neutral Sterol Composition in Bile

Specimens of duodenal bile were obtained by intubation from three symptomatic CTX subjects (J.C., E.D.S., and E.D.E) and three normal volunteers and were analyzed (Table V) (7). Cholesterol accounted for more than 99% of the sterols in the bile from the normal subjects with only very small amounts of cholestanol (0.7%) and a trace of lanosterol (0.02%) detectable. In the symptomatic CTX subjects, the amount of cholesterol in the bile ranged from 85–95% of the total sterols. There was, however, a 10–40-fold increase in the cholestanol content, and substantial amounts of lanosterol, dihydrolanosterol, and 5α-cholest-7-en-3β-ol (Δ^7-cholestenol) were present. The demonstration of the increased proportion of cholestanol in the bile as compared with the plasma of the CTX subjects suggests that the liver preferentially secretes this sterol into the bile. The secretion of substantial quantities of the cholesterol precursors—lanosterol, dihydrolanosterol, and Δ^7-cholestenol—implies that hepatic cholesterol synthesis was hyperactive (see below). The absence of the cholestanol precursors from all other tissues, including brain and xanthomas, is evidence against active sterol biosynthesis by these tissues. Therefore, we assume that the accumulation of cholesterol and cholestanol results from the enhanced deposi-

TABLE V. Neutral Sterol Composition in Bile

	3 normal subjects	Patient J. C.	Patient E. D. E.	Patient E. D. S.
Sterols, mg/ml	1.2	2.0	4.9	4.1
Cholesterol, %	99.28	89.9	95.1	84.8
Cholestanol, %	0.7	6.7	3.7	10.6
Δ^7-Cholestenol, %	—	1.6	1.0	1.2
Lanosterol, %	0.02	1.3	0.2	2.8
Dihydrolanosterol, %	—	0.5	0.01	0.8

tion of these sterols in the tissues from the plasma and that the site of their production is the liver.

E. Bile Acid Composition in Bile

Biliary bile acid composition in three CTX subjects and three normal volunteers was determined by gas–liquid chromatography (GLC) on columns packed with 1% HiEff 8BP. The results are reported in Table VI (7). About 88% of the bile acids from the normal volunteers consisted of cholic acid and chenodeoxycholic acid which were present in approximately equal proportions; the remaining 12% was deoxycholic acid. In the three CTX patients, there was a marked reduction in the proportion of chenodeoxycholic acid and of deoxycholic acid; in patient J.C., no deoxycholic acid was detected. In contrast to the normal subjects, the proportion of cholic acid rose to about 80% of the total bile acids. In addition, 5–9% of the bile acid fraction was present in GLC peaks which were not detected in the normal subjects. These compounds have now been identified conclusively by gas chromatography–mass spectrometry (see below) as 5β-cholestane-3α,7α,12α,25-tetrol, 5β-cholestane-3α,7α,12α,24α,25-pentol, 5β-cho-

TABLE VI. Percentages of Total Bile Acids in Duodenal Biliary Drainage

Acid	3 normal subjects, %	Patient J. C., %	Patient E. D. S., %	Patient E. D. E., %
Cholic acid	44.7 ± 5.0	88	77	86
Chenodeoxycholic acid	43.4 ± 4.9	3	11	6
Deoxycholic acid	11.8 ± 9.6	—	7	2
Allocholic acid	—	4	2.5	4
Unidentified	—	5	2.5	2

lestane-3α,7α,12α,23ξ,25-pentol,   and   5β-cholestane-3α,7α,12α,22ξ,25-pentol. The demonstration of the tetra- and pentahydroxy bile alcohols with a hydroxy group in the 12α-position indicates that these compounds cannot serve as precursors of chenodeoxycholic acid, as had originally been postulated, because of the diminished amount of that bile acid in the bile of CTX subjects. Obviously, 3α,7α,12α-trihydroxy bile alcohols are likely intermediates on the cholic acid biosynthetic pathway. Therefore, we postulate that the deficiency of chenodeoxycholic acid in the bile of CTX subjects results from a preferential diversion of early bile acid precursors such as 7α-hydroxy-4-cholesten-3-one into the cholic acid pathway. The accumulation of C-12 hydroxy bile alcohols with intact side chain in CTX points to impaired cleavage of the cholesterol side chain as the primary biochemical defect (see below).

VI. QUANTITATIVE ASPECTS OF CHOLESTEROL TURNOVER

During the last few years, a number of methods have become available to measure daily cholesterol turnover. The two that have been most widely used include the sterol balance technique as developed by Miettinen *et al.* (23) and Grundy *et al.* (24) and mathematical analysis of compartmental models which was first applied to cholesterol metabolism by Goodman and Noble in 1968 (25,26). We have used both methods to investigate cholesterol turnover in CTX subjects. The basic assumption which underlies the use of these methods is that a "steady state" prevails. This means that the pool of body cholesterol is constant, and consequently losses of cholesterol from the body are replaced completely either by synthesis or absorbed cholesterol. Although small amounts of cholesterol are lost from the skin or degraded to steroid hormones in the gonads and adrenal gland, the most significant loss of cholesterol occurs by biliary excretion into the feces or degradation to bile acids by the liver. New cholesterol enters the body pools to replace losses either as absorbed dietary cholesterol or as newly synthesized cholesterol (hepatic or intestinal synthesis). Therefore, if the diet does not contain cholesterol (as in the studies described here), then the absorption of exogenous cholesterol is *nil* and only newly synthesized cholesterol is available to replace losses. Since the loss of cholesterol through the skin is small, measurements of the daily fecal output of cholesterol and bile acids by the steroid balance method approximates daily input or synthesis. The second (isotope kinetic) method provides a measure of cholesterol turnover by mathematical analysis of compartmental models. After the intravenous injection of a tracer dose of [4-^{14}C]cholesterol, the semilogarithmic plot of plasma cholesterol specific activity vs. time has been shown to decay in a curvilinear fashion. Further, the specific activity decay curves can be divided into at least two exponentials. Mathematical analysis of these exponentials gives estimates of daily production rates (PR_A) and the size of the rapidly turning over pool (M_A).

In the following study on CTX and control subjects, we compared cholesterol production as measured by these two methods. Three CTX subjects and five control individuals (two with normal lipid levels and three with hyperlipidemia) were hospitalized on a metabolic ward and fed a formula diet that was nutritionally adequate but contained virtually no cholesterol. The fat in the diet was supplied by cottonseed oil and contained 100–200 mg/day of plant sterols. After a period of time to allow for attainment of a steady state as noted by constant weight and constant plasma cholesterol levels, all individuals were given [4-^{14}C]cholesterol intravenously and plasma cholesterol specific activity was measured over the ensuing 4–10 weeks. All decay curves could be fitted into a two-pool model and numerical values of production rates (PR_A) were obtained. Simultaneously, measurements of cholesterol and bile acid output were made by the sterol balance method and compared with turnover values (production rates) derived from mathematical analysis of the decay curves. The results are presented in Table VII (27). In two CTX subjects, cholesterol production (PR_A) as estimated by the isotope kinetic method, averaged 1036 mg/day or 18.2 mg/kg/day when the production rate was normalized for the size of the individual. Sterol balance measurements showed that cholesterol turnover averaged 803 mg/day or 14.2 mg/kg/day. Similar measurements in four control subjects were 736 ± 261 mg/day or 11.1 ± 3.6 mg/kg/day by the isotope kinetic method and 674 ± 187 mg/day or 9.7 ± 2.3 mg/kg/day by the sterol balance technique. The 8% difference in cholesterol production between the sterol balance and isotope kinetic methods in the normal individual is expected, and it tends to validate the data derived from those methods. In the CTX subjects, daily cholesterol production was almost twice that of the controls and indicates that cholesterol synthesis is elevated in this condition. However, the value for production by the isotope kinetic method was over 20% higher than the corresponding value obtained by the sterol balance method. The explanation for this difference was not apparent when these studies were performed, but was later shown to result from the substantial excretion of tetra- and pentahydroxy bile alcohols that are found in bile and feces from CTX patients but not in normal individuals (28). The bile alcohol fraction consists predominantly of: 5β-cholestane-3α,7α,12α,25-tetrol, 5β-cholestane-3α,7α,12α,23ξ,25-pentol, and 5β-cholestane-3α,7α,12α,24α,25-pentol which are too polar to be extracted into the sterol fraction with hexane and are partially destroyed by vigorous saponifi-

TABLE VII. Cholesterol Turnover

Diagnosis	Patient	Isotope kinetic method PR_A, mg/day	Isotope kinetic method PR_A, mg/kg/day	Sterol balance method — Endogenous neutral sterols Chromatographic, mg/day	Sterol balance method — Endogenous neutral sterols Isotopic, mg/day	+ Bile acids Chromatographic, mg/day	= Total mg/day	Total mg/kg/day
CTX	E.D.E.	1085	18.4		788 ± 257 (5)[a]	136 ± 46	924	16.0
CTX	J.C.	987	18.0		588 ± 81 (5)	93 ± 27	681	12.4
	Average	1036	18.2		688	115	803	14.2
Normal	G.S.	665	11.3	529 ± 121 (11)[a]		306 ± 73	834	11.4
Normal	J.T.	634	8.7	463 ± 51 (9)		210 ± 48	674	9.2
Type IIa	M.S.	396	7.3	243 ± 21 (12)		116 ± 42	359	6.6
Type IIa	G.W.	922	11.4	474 ± 39 (8)		234 ± 58	708	8.7
Type IIb	B.Mc.	1062	16.6	590 ± 69 (8)		205 ± 35	795	12.4
	Average ±SD	736 ± 261	11.1 ± 3.6	459 ± 131		214 ± 68	674 ± 187	9.7 ± 2.3

[a] ( ) Number of observations ± standard deviations, SD.

cation in an autoclave that is necessary for the analysis of the bile acids by the method of Grundy *et al.* (24). Therefore, substantial amounts of the bile alcohols were lost during the usual workup of feces. Since these compounds are derived from cholesterol, failure to recover them led to an underestimation of cholesterol production by the sterol balance method and consequently explains the difference between cholesterol turnover measured by the sterol balance technique and the production rates calculated by analysis of specific activity decay curves. Recently, Setoguchi *et al.* have devised an alternative method to extract neutral sterols from feces that recovers over 90% of the more polar bile alcohols (28). This method involves the extraction of the feces with ethanol in a Soxhlet apparatus followed by the separation of all polar and nonpolar neutral sterols by partition between ethyl acetate and aqueous ammonia (*p*H 11). The neutral sterols in the ethyl acetate extract are chromatographed on alumina V and the polar bile alcohols are eluted with 10% methanol in ethyl acetate. When the sterol balance data were corrected for the almost 100 mg/day of bile alcohols excreted in the feces by the CTX subjects, the difference between the isotope kinetic method and sterol balance method was reduced to almost 10%—the same difference as in the control subjects. A 10% difference can probably be accounted for by the shorter duration of the isotope kinetic experiments (Table VIII). Samuel and Lieberman and Goodman *et al.* have also noted a 9–14% overestimate of the production rate (PR_A) if only short-term specific-activity curves (less than 20 weeks) are analyzed (29,30). Thus, another explanation for the difference in cholesterol production rate as calculated by the isotope kinetic method is the fact that only short-term specific activity decay curves were examined (6–10

TABLE VIII. Cholesterol and Bile Acid Turnover in Human Subjects Determined by Isotope Kinetic and Chromatographic Balance Methods

Patient	Production rates, mg/day				Chromatographic balance plus bile alcohols	
	Isotope kinetic method[a]	Sterol balance method[a]			Bile alcohols	Total (steroids + bile alcohols)
		Neutral sterols	Bile acids	Total steroids		
E.D.E.	1,085	788	136	924	153	1,077
J.C.	987	588	93	681	80	761
	1,036[b]	688	115	803	117	919
Controls (5)	736[b] ± 261	459 ± 131	215 ± 68	674 ± 187	—[c]	674 ± 187

[a] Salen and Grundy (27).
[b] Average values (± indicates 1 SD).
[c] Not detectable by methods employed.

weeks). Nevertheless, our findings clearly suggest a relationship between the elevated tissue cholesterol concentrations found in the CTX subjects and the high rates of cholesterol production as measured by the sterol balance and isotope kinetic technique. To obtain additional information on this point and to examine the site of cholesterol formation in CTX subjects, measurements of the hepatic activity of 3-hydroxy-3-methylglutaryl CoA (HMG CoA) reductase were made in CTX subjects and 9 normal volunteers (31). It is now well established that the first totally committed precursor of cholesterol is mevalonic acid, which is formed from HMG CoA. The reaction is catalyzed by the microsomal enzyme HMG CoA reductase. In a recent study, we have obtained evidence that the activity of hepatic HMG CoA reductase correlates with the daily production of cholesterol as measured by the sterol balance method. Similar results were obtained in two other CTX subjects where the activity of hepatic HMG CoA reductase was 3-fold greater than the average value found in 9 control subjects. The marked elevation in HMG CoA reductase in the CTX subjects is consistent with high rates of cholesterol production and confirms the quantitative data on cholesterol turnover obtained by the balance and isotope kinetic methods. Further, these findings point to the liver as the site of excessive cholesterol production in this disease and are supported by the presence of the cholesterol precursors—lanosterol, dihydrolanosterol, and Δ^7-cholestenol in the bile (Table V).

VII. MEASUREMENTS OF CHOLESTANOL SYNTHESIS BY THE STEROL BALANCE AND ISOTOPE KINETIC METHODS (27)

In order to study the daily synthesis of cholestanol, a tracer dose of [1,2-³H]cholestanol was administered intravenously as a pulse-label to two CTX and five control individuals. The specific radioactivities of plasma cholestanol were measured biweekly for the next five weeks. Fig. 3 illustrates the specific activity vs. time curve for [1,2-³H]cholestanol from patient J.C. with CTX. A two-pool model is suggested by the curve-peeling technique of Goodman and Noble (25). Also plotted are the specific activities of cholestanol isolated from the feces of this subject (open circles). It is noteworthy that these values corresponded closely with the specific activities of plasma cholestanol which indicated that newly synthesized, unlabeled cholestanol was not produced by intestinal bacteria.

In two CTX subjects and one normolipidemic subject, cholestanol turnover was calculated by the isotopic balance technique after [1,2-³H]cho-

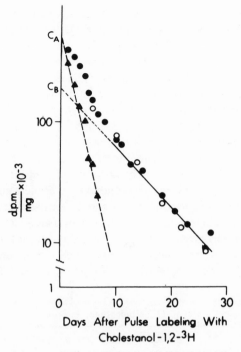

Fig. 3. Specific activity vs. time curve for plasma cholestanol after pulse labeling intravenously with [1,2-³H] cholestanol. A two-pool model is suggested by the curve-peeling technique: the difference values (solid triangles obtained by subtracting the points on the extrapolated line from the experimental points) fit a straight line. ●, specific activity cholestanol plasma; ○, specific activity cholestanol feces; ▲, specific activity difference between extrapolated and experimental points.

lestanol was administered intravenously. Since the specific activity of plasma and fecal cholestanol are equal, it is possible to calculate the milligrams of cholestanol excreted daily in the feces by measuring the 24 h output of tritium in the feces divided by the specific activity of plasma cholestanol. The daily excretion of cholestanol as 5α-cholanoic acids (allo bile acids) can also be computed by assuming that the specific activity of the fecal allo bile acids is the same as that of plasma cholestanol, measuring the 24 h output of tritium in the acidic fraction of stools, and dividing this quantity by the specific activity of plasma cholestanol. Thus, the daily excretion (mg/day) of neutral and acidic steroids derived from cholestanol can be calculated and is presented in Table IX. Cholestanol turnover can also be estimated by the chromatographic balance technique. In this procedure, the daily fecal excretion of cholestanol is measured directly by GLC while the daily excretion of acidic steroids derived from cholestanol is estimated by the following equation: acidic steroids from cholestanol (mg/day) = neutral steroids from cholestanol (mg/day) × total acidic steroids (mg/day) ÷ total endogenous neutral steroids (mg/day). This equation is based on the assumption that the ratio of endogenous cholesterol to bile acids is identical to the ratio of cholestanol to allo bile acids in the feces. It is valid also because over 97% of the tritium radioactivity recovered after TLC of fecal neutral steroids was associated with cholestanol and indicates that cholestanol was not modified further by intestinal bacteria. The values for the excretion of cholestanol are reported in Table IX. The daily fecal output of cholestanol ranged from 22–38 mg/day in the CTX subjects and from 7–13 mg/day in the control subjects. The output of acidic steroids derived from cholestanol ranged from 6–9 mg/day in the CTX subjects and

TABLE IX. Cholestanol Turnover

| Patient | Disease | Cholestanol balance data, mg/day | | | PR_A, mg/day |
		Neutral steroids	Acidic steroids	Total steroids	
E.D.E.	CTX	36	6	42^a	57
J.C.	CTX	38	8	46^a	48
E.D.S.	CTX	22	9	31^b	—
				39.7 ± 8.4^c	52.5
G.S.	N	9	5	14^a	8.0
G.W.	IIa	7	4	11^b	
I.G.	IIa	9	2	11^b	
J.C.	IIb	13	4	17^b	
H.T.	IIb	—	—	—	6.0
R.P.	IV	—	—	—	18.6
R.M.	N	—	—	—	18.0
E.S.	IIa	—	—	—	8.2
				13.3 ± 2.9	11.8 ± 6.0

[a] Neutral and acidic steroids determined by isotope balance method.
[b] Neutral steroids determined by chromatographic balance method. Acidic steroids were calculated indirectly by assuming that the percentage of bile acids derived from cholestanol was identical to the percentage of bile acids derived from cholesterol. This assumption was proved to be valid in patients J.C. and E.D.E. in whom acidic steroids derived from cholestanol were measured directly by isotopic means. Thus, the acidic steroids in this table were calculated according to the following equation: acidic steroids (from cholestanol) (mg/day) = neutral steroids (from cholestanol) (mg/day) × total acidic steroids (mg/day) ÷ endogenous neutral steroids (mg/day).
[c] Means ±SD.

TABLE X. Half-Life of [1,2-³H]Cholestanol: Exponentials A and B and Size of Pool A and Pool B

Patient	C_A, dpm/mg[a]	C_B, dpm/mg[a]	$t\frac{1}{2}A$, days	$t\frac{1}{2}B$, days	M_A, mg	M_B,[b] mg
E.D.E.	350,000	66,000	1.7	7.0	212	212
J.C.	267,000	152,000	1.8	5.8	148	67
		Av.	1.8	6.4	180	140
G.S.	940,000	290,000	1.6	12.0	48	67
H.T.	1,000,000	870,000	2.0	5.4	31	9
R.P.	960,000	125,000	1.1	5.8	54	57
R.M.	460,000	240,000	1.0	6.0	83	62
E.S.	3,000,000	620,000	1.0	6.0	22	23
		Mean $\pm$ SD	1.3 ± 0.4	7.0 ± 2.8	48 ± 23	43 ± 26

[a] Constants which represent intercept of the exponential lines A and B with the y axis when $t = 0$.
[b] Minimum value.

from 2–5 mg/day in the control subjects. Since these individuals had attained a metabolic steady state as manifested by constant body weight and plasma cholesterol and cholestanol concentrations, the sum of the fecal excretion of cholestanol and allo bile acids derived from cholestanol reflects daily turnover.

In two CTX subjects (J.C. and E.D.E.) and five other subjects, cholestanol turnover was estimated through the calculation of production rates (PR_A) by mathematical analysis of the specific activity decay curves, since all curves conformed to two-pool models. The various kinetic parameters that underlie these calculations are shown in Table X and the production rates (PR_A) are included in Table IX. Cholestanol turnover (PR_A) ranged from 6.0–18.6 mg/day in the control subjects (mean 11.8 ± 6.0) and 48 and 57 mg/day (average 52.5) respectively in the two CTX subjects. Thus, average cholestanol PR_A values were 3–4 times greater in CTX subjects than the controls. Since there was virtually no cholestanol present in the diets fed to these subjects, daily turnover was equal to the amount of cholestanol synthesized each day when the individuals were in the steady state. Therefore, cholestanol synthesis in CTX individuals is greatly increased over normal and probably accounts for the elevated tissue and plasma concentrations of cholestanol that characterize this condition. Additional evidence of excessive body cholestanol accumulation was obtained from calculations of the size of the two miscible cholestanol pools (M_A and M_B). The calculations were made by the formula given by Goodman and Noble (25), and the results are listed in Table X. The size of pool A (M_A), the rapidly turning-over pool of cholestanol in two CTX subjects, was 212 mg and 148 mg respectively; both values, as expected, exceeded the mean value for Pool A (M_A = 48 + 23 mg) in the control subjects by 4 and 3 times respectively. The tissues which compose pool A are red cells, liver, and intestine, and the cholestanol in those tissues equilibrated rapidly with plasma lipoprotein cholestanol. The size of cholestanol Pool B (M_B) was 212 mg and 67 mg in the two CTX subjects as compared with a mean value of 43 ± 26 mg in 5 control subjects. Although the accuracy of the calculation for M_B is still questioned it is probably a minimum estimate because we have assumed that neither synthesis or degradation of cholestanol occurs in the pool. The values for the CTX subjects are clearly elevated and support the quantitative measurements of increased tissue concentrations.

VIII. PLASMA LIPOPROTEIN TRANSPORT AND ESTERIFICATION OF CHOLESTANOL (27)

To determine the lipoprotein fraction which transports cholestanol, plasma specimens from two CTX individuals were fractionated by preparative ultracentrifugation into the three major lipoprotein groups (VLDL, LDL, and HDL). The absolute amounts of cholesterol and cholestanol then were determined by gas–liquid chromatography in each lipoprotein fraction and are reported in Table XI. Cholestanol accompanied cholesterol in all three lipoprotein classes. In one subject, the ratio of cholesterol to cholestanol was reasonably constant throughout the three lipoprotein fractions but was higher in the VLDL and HDL fractions of the other subject. However, this experiment clearly shows that cholestanol is transported with cholesterol and that probably no special lipoprotein for cholestanol transport exists.

The rate of cholestanol esterification was compared with that of cholesterol esterification following the simultaneous administration of [1,2-

TABLE XI. Cholesterol and Cholestanol in Plasma Lipoproteins

	Cholesterol/cholestanol, mg/mg	
	Patient 1	Patient 3
Plasma (per 100 ml)	145/1.7	151/2.4
VLDL	28/0.22	18/0.29
LDL	106/1.5	113/1.9
HDL	11/0.07	20/0.42

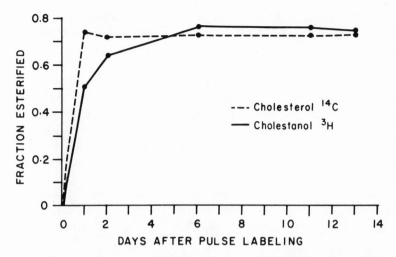

Fig. 4. The esterification of [4-¹⁴C]cholesterol and [1,2-³H]cholestanol. The relative
rates of esterification of both isotopic sterols were compared in one CTX patient.

³H]cholestanol and [4-¹⁴C]cholesterol to two CTX subjects and one normal
control. Plasma specimens were obtained over the next 13–19 days and the
distribution of the isotopes between free and ester fractions was measured.
The results for one subject are given in Fig. 4. In all three subjects, 70–75%
of radioactive cholesterol and cholestanol were esterified after several days
and this was confirmed by mass measurements. Apparently, trace quantities
of the two radioactive sterols were treated identically by the body. In two
subjects, cholestanol was esterified at a somewhat slower rate than cholesterol, but as time progressed, slightly more cholestanol was in ester form than
cholesterol. This experiment suggests that the synthesis and catabolism of
cholestanol esters are similar to cholesterol ester, but the rates for cholestanol are slightly lower.

IX. CHOLESTANOL BIOSYNTHESIS (32)

Although much is known about the biosynthesis of cholesterol, very
little information is available about cholestanol biosynthesis. Neither the
tissue of synthesis nor the biosynthetic pathway has been precisely elucidated. In order to gather information on this subject, two independent
studies were conducted in CTX patients who showed augmented cholestanol
production. The first study dealt with conversion of [4-¹⁴C]cholesterol into

cholestanol. After a tracer dose of radioactive cholesterol was given intravenously to a CTX subject, neutral sterols (cholesterol, cholestanol, dihydrolanosterol, and lanosterol) were isolated from the bile over the next week and the specific activities compared. The results are shown in Fig. 5. The presence of label in cholestanol proved that cholesterol was converted into cholestanol. After isolation of the cholestanol by AgNO₃-TLC, peroxyformic acid oxidation was performed to eliminate traces of cholesterol that might have migrated with cholestanol on TLC. Although it is unlikely that [4-^{14}C]cholesterol was converted to its precursor, Δ^7-cholestenol, had any been present, the oxidation procedure also would remove this sterol.

The demonstration of increasing specific radioactivity of [4-^{14}C]cholestanol during the 5-day study coupled with the finding that the cholestanol specific activity curve intersected the cholesterol specific activity about day

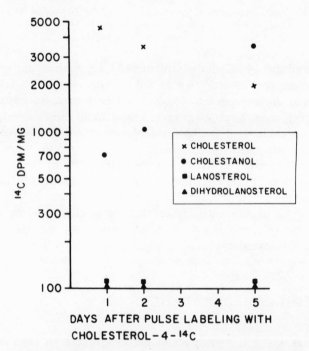

Fig. 5. Specific activity vs. time curves for [4-^{14}C]cholesterol and [4-^{14}C]cholestanol isolated from bile after patient J.C. received 30 μCi of [4-^{14}C]cholesterol intravenously. The specific activity of [4-^{14}C]cholesterol decreased while the specific activity of [4-^{14}C]cholestanol increased and exceeded [4-^{14}C]cholesterol by the 5th day.

TABLE XII. Isotope Ratios (^{3}H/^{14}C) of Sterols Isolated after Incubation of [2-^{14}C]-Mevalonate and 3R,4R,-[4-^{3}H]Mevalonate with Rat Liver Homogenates

| | | | ^{3}H/^{14}C | | |
Standards	^{3}H, dpm	^{14}C, dpm	Actual	Relative to squalene	Predicted
Squalene	29,700	6,200	4.79	1.00	—
Lanosterol[a]	9,800	2,400	4.08	0.85	0.83
Dihydrolanosterol[a]	15,900	3,800	4.18	0.87	0.83
Cholesterol	62,500	20,100	3.11	0.65	0.60

[a] Purification facilitated by the addition of chemically pure, nonradioactive sterol before AgNo$_3$-TLC.

4 suggests a precursor product relationship between cholesterol and cholestanol. Had cholestanol arisen from the reversal of the cholesterol synthetic pathway, label would be expected in the cholesterol precursors lanosterol and dihydrolanosterol. Since significant counts were not found in these compounds, this possibility is considered excluded.

In a second experiment, the biosynthesis of cholestanol and cholesterol from stereospecifically labeled mevalonate was considered. A CTX subject received a mixture of [2-^{14}C]mevalonate (200 μCi) and 3R,4R-[4-^{3}H]mevalonate (1000 μCi) as a single intravenous pulse-label. This experiment was suggested by Dr. George Popják and is based on the work of Cornforth, Popják, and colleagues (33) who, by means of degradative studies, established the precise locations of the ^{14}C and ^{3}H atoms in squalene, lanosterol, and cholesterol synthesized by liver homogenates from the mevalonate mixture (Table XII). This type of experiment can be used to find out whether lanosterol or Δ^7-cholestenol, two saturated sterols belonging to the 5α-cholestane series, and which are both precursors of cholesterol, could be transformed directly into 5α-cholestanol without involvement of cholesterol, which has a double bond at C-5,6. The theoretical consideration which underlies this experiment is illustrated in Fig. 6. This figure shows that six molecules of mevalonate are incorporated into each molecule of squalene. Since a mixture of [2-^{14}C] and 3R,4R-[4-^{3}H] mevalonate were administered to the subject, the ^{3}H/^{14}C ratio in the squalene is 6/6 or 1; while the ^{3}H/^{14}C ratio in lanosterol is 5/6 (a ^{3}H atom is lost from C-7 during the cyclization of the squalene to lanosterol). Further rearrangements and losses of the radioactive atoms occur as lanosterol is transformed to cholesterol to yield a final ^{3}H/^{14}C ratio of 3/5. If cholestanol was formed from cholesterol, then the ^{3}H/^{14}C ratio will be 3/5, or the same as in cholesterol. However, if lanosterol (^{3}H/^{14}C = 5/6) or Δ^7-

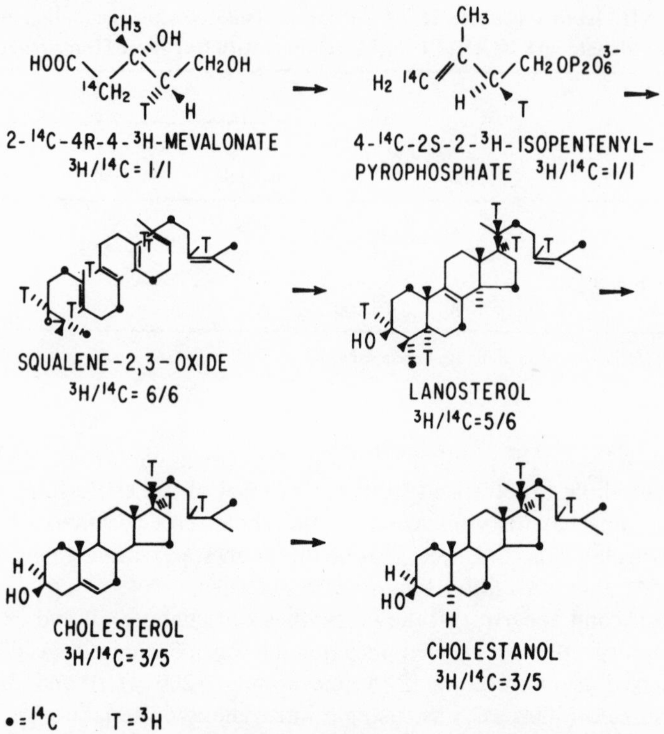

2-^{14}C-4R-4-^{3}H-MEVALONATE
^{3}H/^{14}C = 1/1

4-^{14}C-2S-2-^{3}H-ISOPENTENYL-
PYROPHOSPHATE ^{3}H/^{14}C=1/1

SQUALENE-2,3-OXIDE
^{3}H/^{14}C = 6/6

LANOSTEROL
^{3}H/^{14}C=5/6

CHOLESTEROL
^{3}H/^{14}C = 3/5

CHOLESTANOL
^{3}H/^{14}C=3/5

● = ^{14}C   T = ^{3}H

Fig. 6. Stereospecific labeling of squalene, lanosterol, and cholesterol based on previous studies in lower species, after administration of [2-^{14}C-4R,4-^{3}H]mevalonate and postulated distribution of radioactive atoms in 5α-cholestanol. Experimental finding of a 3/5 ratio of ^{3}H/^{14}C in 5α-cholestanol relative to squalene indicates that cholesterol (^{3}H/^{14}C = 3/5) served as the direct precursor.

cholestenol (^{3}H/^{14}C = 4/5) were converted directly into cholestanol bypassing cholesterol, then the ^{3}H/^{14}C ratio in cholestanol will be at least 20% greater than cholesterol or 4/5 because of the presence of the additional ^{3}H atom at C-5 derived directly from the lanosterol and Δ^{7}-cholestenol. In a preliminary experiment, a portion of the stereospecifically labeled mixture of mevalonate was incubated anaerobically with rat liver homogenate to give squalene and then aerobically to yield lanosterol, dihydrolanosterol, and cholesterol. The results of this experiment (performed by Drs. Alan Polito and George Popják) (32) are presented in Table XII. It is assumed that the ratio of ^{3}H/^{14}C in squalene is 6/6 or 1. The experimental ratios found in lanosterol, dihydrolanosterol, and cholesterol relative to squalene were almost identical to the predicted values. Similar results were obtained from the *in vivo* experiments carried out in a CTX subject (Table XIII). The

TABLE XIII. Isotope Ratios (^{3}H/^{14}C) and Specific Activities of Sterols Isolated from the Bile of a CTX Patient[a]

Sterols	Days after pulse labeling	Mass, mg	^{3}H, dpm	^{14}C, dpm	^{3}H/^{14}C		Specific activities, dpm/mg	
					Actual	Relative to squalene	^{3}H	^{14}C
Cholesterol	1	30.10	1,380,700	449,600	3.07	0.64	45,800	14,900
	3	29.10	479,200	156,800	3.06	0.64	16,400	5,380
	5	34.90	427,600	139,300	3.07	0.64	12,300	4,000
Cholestanol + Δ^7-cholestenol	1	1.70	106,300	15,900	6.69	1.40	62,400	9,340
	3	2.50	18,100	3,960	4.57	0.95	7,200	1,570
	5	4.50	25,700	6,190	4.15	0.87	5,940	1,430
Lanosterol	1	0.25	6,100	1,390	4.39	0.92	24,600	5,610
	3	0.72	1,230	280	4.39	0.92	1,710	390
	5	1.13	1,190	270	4.41	0.92	1,050	240
Dihydrolanosterol	1	0.04	1,760	440	4.00	0.84	48,100	11,900
	3	0.15	850	250	3.40	0.71	5,700	1,660
	5	0.34	1,770	540	3.28	0.68	5,220	1,600

[a] Patient J.C.

sterols were isolated by argentation TLC from three specimens of bile obtained over a period of five days, and the $^3H/^{14}C$ ratios are expressed relative to squalene. Good agreement between the predicted and experimental $^3H/^{14}C$ ratios were found for cholesterol, lanosterol, and dihydrolanosterol. The values for cholestanol differed from any predicted values since the $^3H/^{14}C$ ratio of 1.40 for the cholestanol + Δ^7-cholestenol fraction on day one suggests the presence of additional contaminants. It would be highly unlikely that the squalene biosynthesized from the mevalonate mixture would give rise to a sterol with a $^3H/^{14}C$ ratio greater than 1.0. Therefore, to eliminate this contaminant and remove any Δ^7-cholestenol which is not separated from cholestanol by argentation thin layer chromatography, the mixture was subjected to peroxyformic acid oxidation. This technique removes unsaturated compounds but does not affect saturated sterols like cholestanol. The cholestanol can be separated from the more polar oxidized compounds by solvent partition. The results of this oxidation experiment are shown in Table XIV. After peroxyformic acid exposure, the contaminants were removed and the true $^3H/^{14}C$ ratio in the cholestanol relative to squalene was almost identical to that of cholesterol. This experiment conclusively proves that cholestanol was derived from cholesterol and not from cholesterol precursors with the 5α-configuration.

Thus, two independent experiments show that cholesterol is the precursor of cholestanol. The precise pathway has not been elucidated. However, Rosenfeld and colleagues showed that $[3\alpha\text{-}^3H,4\text{-}^{14}C]$cholesterol was converted to cholestanol with the almost total elimination of the $3\alpha\text{-}^3H$ atom (34). Those investigators interpret their results as showing the following pathway: Cholesterol $\rightarrow$ 4-cholesten-3-one $\rightarrow$ 5α-cholesten-3-one $\rightarrow$ 5α-cholestane-3β-ol. Thus, the reduction of the double bond of cholesterol from cholestanol involves a 3-oxo intermediate. Indeed, 4-cholesten-3-one is rapidly converted into cholestanol when given either to normal individuals or CTX subjects. However, information on the formation of 4-cholesten-3-one from cholesterol is scanty. Shefer and colleagues have shown that 0.1–

TABLE XIV. Isotope Ratios ($^3H/^{14}C$) in Cholestanol after Peroxyformic Acid Oxidation

Days after pulse labeling	Mass, mg	^3H, dpm	^{14}C, dpm	$^3H/^{14}C$		Specific activities, dpm/mg	
				Actual	Relative to squalene	^3H	^{14}C
1	0.20	1600	490	3.26	0.68	8000	2500
3	0.12	600	200	3.00	0.63	5200	1700
5	0.49	3700	1200	3.08	0.64	7600	2500

0.5% of [4-^{14}C]cholesterol incubated with rat liver homogenates is transformed to 4-cholesten-3-one (35). Similar findings have also been reported by Björkhem and Karlmar who incubated radioactive cholesterol with rat liver microsomes and detected 4-cholesten-3-one (36).

We have compared the transformation of cholesterol into 4-cholesten-3-one by normal and CTX liver. Microsomal preparations from the liver of a CTX patient had about 10 times the capacity of producing 4-cholesten-3-one as normal liver. These findings are consistent with the elevated quantitative measurements of cholestanol production in CTX and strongly imply that cholestanol production occurs in the liver. Since bile acid production is abnormal in CTX (see below), we attempted to show a relationship between a known bile acid intermediate, 7α-hydroxy-4-cholesten-3-one and 4-cholesten-3-one which is the key intermediate in the cholestanol biosynthetic pathway (37). A mixture of [G-^{3}H]7α-hydroxy-4-cholesten-3-one and [4-^{14}C]4-cholesten-3-one was given intravenously to a CTX subject and a normal individual and radioactivity in biliary bile acids and cholestanol was determined. Both the normal control and the CTX individual responded similarly: ^{3}H was detected only in the bile acid fraction (cholic acid, chenodeoxycholic acid, and deoxycholic acid) and was not present in the cholesterol or cholestanol fraction. This clearly indicated that 7α-hydroxy-4-cholesten-3-one which is formed from cholesterol via 7α-hydroxycholesterol and is an intermediate in the bile acid biosynthetic pathway is not the source of cholestanol. In contrast, ^{14}C was found in cholestanol and small amounts in the allo bile acids. Thus, 4-cholesten-3-one was shown to be converted into cholestanol which, in turn, served as a precursor of the 5α-series of bile acids. However, this study clearly showed that 7α-hydroxy-4-cholesten-3-one is not dehydroxylated to yield 4-cholesten-3-one. Therefore, the synthesis of cholestanol and of bile acid occurs via specific pathways and intermediates are not shared. At the present time, we are unable to explain the increased formation of cholestanol in CTX, but it may be assumed that it is due to the increased activity of enzymes catalyzing the formation of 4-cholesten-3-one from cholesterol. This might involve a 3β-hydroxysteroid dehydrogenase and a $\Delta^{5,6} \rightarrow ^{4,5}$ isomerase but the exact sequence of these reactions is not known. Once 4-cholesten-3-one is produced it can be rapidly converted to cholestanol. The following experimental evidence supports this hypothesis: (a) When CTX subjects are given the bile acid sequestering agent, cholestyramine, there is a marked rise in the plasma cholestanol concentration and increased output of cholestanol in the feces which suggests increased formation. Since cholestyramine binds intestinal bile acids and promotes bile acid synthesis by activating the bile acid synthetic pathway and increases the conversion of cholesterol to bile acid, we believe that a portion of the cholesterol was transformed to 4-cholesten-3-one and then to cholestanol. (b) Conversely, when chenodeoxycholic acid is given to CTX individuals there is a marked reduction in bile acid production, since presumably cholesterol synthesis is inhibited by the administration of this bile acid. Cholestanol synthesis and plasma cholestanol levels were reduced fourfold during treatment with this bile acid and this suggests that chenodeoxycholic acid inhibited the formation of 4-cholesten-3-one as well as of cholestanol (38).

X. DEFECTIVE BILE ACID SYNTHESIS IN CTX

We have alluded to a defect in bile acid synthesis in CTX throughout this review. The evidence in favor of this hypothesis can be summarized as follows: In sharp contrast to the elevation of cholesterol biosynthesis in CTX, bile acid production as measured by the sterol balance technique was subnormal (27). In two CTX patients, bile acid excretion averaged 114 mg/day, which was about 50% of the average value of 214 mg/day found in five hyperlipidemic control subjects. Paradoxically, the activity of cholesterol 7α-hydroxylase, the key rate-determining enzyme for bile acid synthesis, was significantly elevated: 30 units in a CTX patient, compared with 20 units in the control group, and 10 units in gallstone subjects (39). Since the rate-determining step in bile acid synthesis is the rate of formation of 7α-hydroxycholesterol from cholesterol, CTX patients should be able to produce adequate amounts of 7α-hydroxycholesterol for normal bile acid production. However, since bile acid production as measured by the sterol balance method was abnormally low, we suspected the existence of an enzyme defect which was responsible for the incomplete transformation of 7α-hydroxycholesterol to the bile acids. This assumption proved correct: more than 10%, about 100 mg/day, of the neutral sterol fraction isolated from the feces of CTX patients was composed of compounds more polar than cholesterol (28). These polar neutral sterols were not present in bile and feces of normal subjects. The polar sterols were extracted and separated chromatographically into two fractions: the first consisted almost entirely of a tetrahydroxy bile alcohol, 5β-cholestane-$3\alpha,7\alpha,12\alpha,25$-tetrol; the second was a complex mixture of pentahydroxy bile alcohols in which 5β-cholestane-$3\alpha,7\alpha12\alpha,24\alpha,25$-pentol and 5β-cholestane-$3\alpha,7\alpha,12\alpha,23\xi,25$-pentol predominated.

The discovery of a family of C-25 hydroxylated bile alcohols was surprising: we had expected to find C-26 hydroxylated intermediates, since the known pathway of bile acid synthesis involves the 26-hydroxylation of bile alcohols (Fig. 7). For example, in the biosynthesis of cholic acid (X)

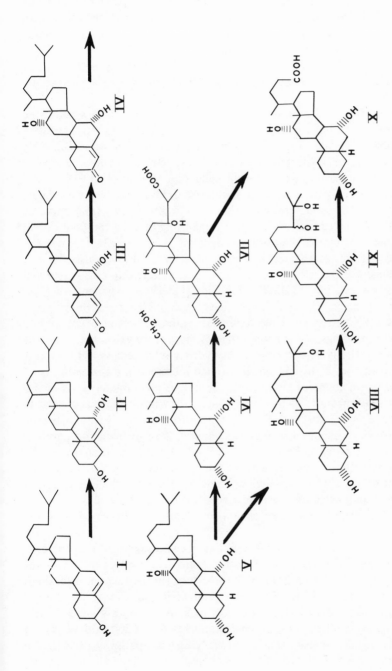

Fig. 7. Pathway of cholic acid biosynthesis showing side chain degradation *via* both 26-hydroxylated and alternatively 25-hydroxylated intermediates. I, cholesterol; II, cholest-5-ene-3β,7α-diol; III, 7α-hydroxycholest-4-en-3-one; IV, 7α,12α,-dihydroxycholest-4-en-3-one; V, 5β-cholestane-3α,7α,12α-triol; VI, 5β-cholestane-3α,7α,12α,26-tetrol; VII, 3α,7α,12α,24ξ-tetrahydroxy-5β-cholestanoic acid; VIII, 5β-cholestane-3α,7α,12α,25-tetrol; IX, 5β-cholestane-3α,7α,12α,24ξ,25-pentol; and X, cholic acid.

5β-cholestane-3α,7α,12α-triol (V) is presumably transformed into 5β-choles-
tane-3α,7α,12α,26-tetrol (VI). We, therefore, considered the possibility
that in CTX, there exists an alternate pathway of cholic acid biosynthesis.
This pathway follows the known steps from cholesterol (I) to 5β-cholestane-
3α,7α,12α-triol (V), and then continues via 5β-cholestane-3α,7α,12α,25-
tetrol (VIII) and 5β-cholestane-3α,7α,12α,24ξ,25-pentol (IX) to yield cholic
acid (Fig. 7).

To substantiate the existence of this pathway, we have studied the
metabolism of tritium-labeled 5β-cholestane-3α,7α,12α,25-tetrol in two
CTX patients (40). The results of one of the studies are shown in Fig. 8. A
tracer dose of tetrol was injected intravenously; tetrol and cholic acid were
isolated from successive two-day stool collections and purified to constant
specific radioactivity. Label from the tetrol (closed circles) was incorporated into cholic acid (crosses). The specific activity vs. time curve of
cholic acid intersected that of the tetrol, suggesting that the cholestanetetrol
served as the precursor of the primary bile acid.

When tritium-labeled tetrol was administered intravenously to two
normal subjects, radioactivity was incorporated rapidly into cholic acid and
the semilogarithmic plot of the specific activity decayed exponentially (Fig.
9). In these control subjects, the turnover of cholic acid was more rapid
than in the CTX patient (in whom bile acid synthesis was impaired). On the
basis of these findings, we conclude that in man, there exists a pathway of
bile acid biosynthesis that involves C-25 hydroxy intermediates. However, it
is not clear whether 25-hydroxylation in man represents a major pathway of
cholic acid or whether side chain oxidation occurs predominantly via the C-
26 hydroxy intermediates as has been suggested by most previous studies. It
is also conceivable that in CTX the 26-hydroxylation of 5β-cholestane-
3α,7α,12α-triol is impaired so that the alternate 25-hydroxylation pathway
becomes the preferred pathway. Since chenodeoxycholic acid appears to be
formed preferentially from 26-hydroxylated intermediates, 5β-cholestane-
3α,7α,26-triol is rapidly 12α-hydroxylated to form 5β-cholestane-3α,7α,
12α,26-tetrol and, eventually, cholic acid—this hypothesis would also serve
to explain the virtual absence of chenodeoxycholic acid in the bile of CTX
patients.

Recent work in other laboratories has likewise suggested that the degradation of the cholesterol side chain to cholic acid involves intermediates
hydroxylated at C-25. Cronholm and Johansson (in the rat) and Björkhem
et al. (in man) reported that the major product formed during the incubation of 5β-cholestene-3α,7α12α-triol with liver microsomes was 5β-choles-
tane-3α,7α,12α,25-tetrol, not 5β-cholestane-3α,7α,12α,26-tetrol as would
be expected from a consideration of the "classical" pathway (41,42). We
have recently demonstrated that 5β-cholestane-3α,7α,12α-triol can be

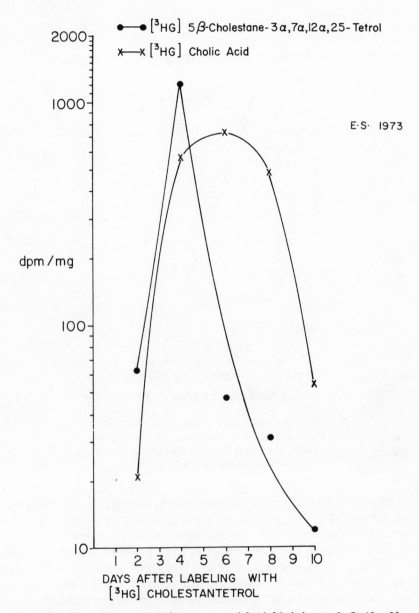

Fig. 8. Specific activity vs. time decay curves of fecal 5β-cholestane-3α,7α,12α, 25-tetrol and cholic acid isolated from a CTX subject after intravenous pulse-labeling with [G-³H]5β-cholestane-3α,7α,12α,25-tetrol.

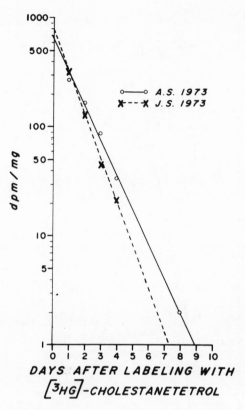

Fig. 9. Specific activity vs. time decay curves of cholic acid from bile of two normolipidemic individuals after intravenous pulse-labeling with [G-³H]5β-cholestane-3α,7α,12α,25-tetrol.

converted to cholic acid in human and rat liver *in vitro,* via a 25-hydroxylation pathway (Fig. 10), without the participation of 26-oxygenated derivatives (43). Incubation of 5β-cholestane-3α,7α,12α,25-tetrol with the microsomal fraction under the conditions employed resulted in the formation of a mixture of 5β-cholestanepentols. When microsomes prepared from human liver (from either normolipidemic controls or CTX patients) were used, 5β-cholestane-3α,7α,12α,24β,25-pentol was the predominant component, while with rat liver microsomes 5β-cholestane-3α,7α,12α,24α, 25-pentol and 5β-cholestane-3α,7α,12α,24β,25-pentol were formed in equal amounts and constituted the major products formed (Table XV).

The 25-hydroxylation pathway of cholic acid biosynthesis, just like the one proceeding via 26-hydroxylated intermediates involves the introduction

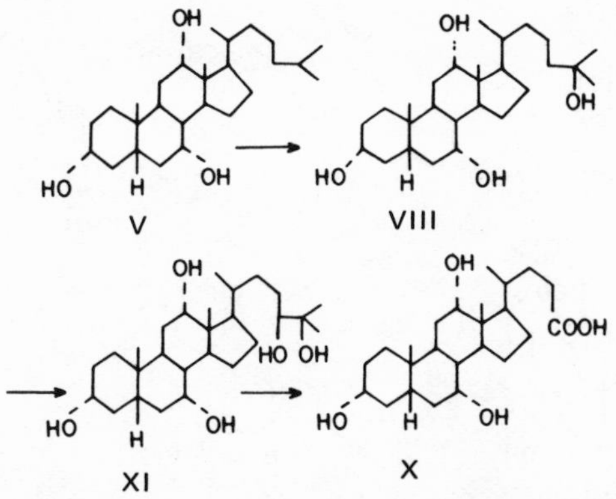

Fig. 10. Pathway of cholic acid biosynthesis showing side chain degradation via 25-hydroxylated intermediates. V, 5β-cholestane-3α,7α,12α-triol; VIII, 5β-cholestane-3α,7α,12α,25-tetrol; XI, 5β-cholestane-3α,7α,12α,24β,25-pentol; X, cholic acid.

of a hydroxyl group at C-24. (Figs. 10,11) It seems, however, that the enzymes which carry out the 24-hydroxylation are not similar for the two pathways, suggesting that different mechanisms are involved. In the 25-hydroxylation pathway, the reaction is catalyzed solely by a microsomal enzyme which was not stimulated by the addition of 100,000 g supernatant solution. Mitochondria and 100,000 g supernatant solution, either alone or in combination, did not catalyze the formation of 24-hydroxylated 5β-cholestane-pentols (Table XVI). In contrast, Masui and Staple, and more recently, Gustafsson found that the introduction of the C-24 hydroxyl group into 3α,7α,12α-trihydroxy-5β-cholestan-26-oic acid (XII, Fig. 11) was catalyzed

TABLE XV. Composition of Pentol Fraction Following Incubation of 5β-Cholestane-3α,7α,12α,25-tetrol with Hepatic Microsomes[a]

	% of total pentol fraction in 5β-cholestane-			
Species	3α,7α,12α,23ξ,25-pentol	3α,7α,12α,24α,25-pentol	3α,7α,12α,24β,25-pentol	3α,7α,12α,25,26-pentol[b]
Human control	23.4	10.5	40.7	25.4
Human CTX	18.2	10.2	44.5	27.3
Rat	17.0	35.6	36.4	11.0

[a] Microsomal fractions were prepared and products were analyzed as described (43).
[b] 5β-Cholestane-3α,7α,12α,25,26-pentol was most probably of microsomal origin, since the microsomal fraction was minimally contaminated with mitochondria (43).

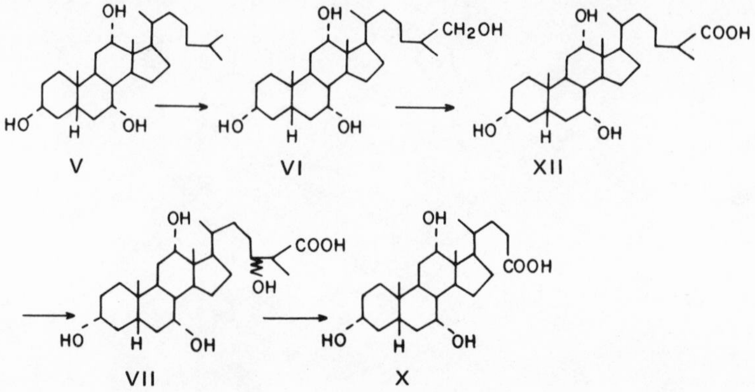

Fig. 11. Pathway of cholic acid biosynthesis showing side chain degradation via 26-hydroxylated intermediates. V, 5β-cholestane-3α,7α,12α-triol; VI, 5β-cholestane-3α,7α,12α,26-tetrol; XII, 3α,7α,12α-trihydroxy-5β-cholestan-26-oic acid; VII, 3α,7α,12α,24ξ-tetrahydroxy-5β-cholestan-26-oic acid; X, cholic acid.

either by the mitochondrial fraction (44,45) or by the microsomal fraction fortified with 100,000 g supernatant solution. In the case of the microsomal fraction, large amounts of ATP were required (45). Moreover, the 24-hydroxylation of 5β-cholestane-3α,7α,12α,25-tetrol (VIII, Fig. 10) requires O_2 and NADPH (Table XVII) and is probably catalyzed by a mixed function oxidase. The studies by Gustafsson (45) postulated that in the pathway involving 26-oxygenated intermediates, the 24-hydroxylation step was not catalyzed by a mixed function oxidase, but was analogous to fatty acid β-oxidation and involved an acyl dehydrogenase and enoyl hydrase as had also been suggested by Masui and Staple (44).

In the present studies, the rate of 5β-cholestanepentol formation was considerably higher in rat than in human liver microsomes, hydroxylation was about four times greater in the human control subjects than in the CTX patients (Table XVI). This reduced rate of hydroxylation in CTX patients might possibly explain the accumulation of 5β-cholestane-3α,7α,12α,25-tetrol in bile and feces of the latter (28).

The enzyme catalyzing the transformation of 5β-cholestane-3α,7α,12α,24,25-pentol to cholic acid was located predominantly in the 100,000 g supernatant fraction (Table XVIII). This transformation proceeded at approximately equal rates in man and in the rat. In both species the reaction appeared to be quite stereospecific in that only the 24β-epimer was transformed to cholic acid at an appreciable rate (Table XIX). This stereospecificity and the rapid metabolism of the 24β-epimer to cholic acid

TABLE XVI. Subcellular Distribution of Side-Chain Hydroxylation System Acting upon 5β-Cholestane-3α,7α,12α,25-tetrol[a]

	Rate of production of 5β-cholestanepentols, pmol/mg protein/10 min		
	Human liver		
Fraction	Control	CTX	Rat liver
Whole homogenate	317	91	2,690
Intact mitochondria[b]	10	4	73
Microsomes	1,458	345	11,680
100,000g supernatant	436	105	70
Mitochondria + 100,000g supernatant	—[c]	—[c]	65
Microsomes + 100,000g supernatant	—[c]	—[c]	3,630

[a] Subcellular fractionation was performed as described (43).
[b] When 5β-cholestane-3α,7α,12α,25-tetrol was incubated with partially broken rat liver mitochondria the rate of hydroxylation increased to 950 pmoles/mg protein/10 min which was less than 10% of the rate observed with the microsomal fraction.
[c] Not analyzed.

might explain the fact that only 5β-cholestane-3α,7α,12α,23ξ,25-pentol and 5β-cholestane-3α,7α,12α,24α,25-pentol could be detected in bile and feces of CTX patients. We were unable to find 5β-cholestane-3α,7α,12α,24β,25-pentol, presumably because it was rapidly transformed into cholic acid (46). The soluble enzyme system required NAD for the formation of cholic acid (Table XX). The reaction might involve a 24-ketone intermediate, which was, however, not observed in the present study. Until further work has

TABLE XVII. Cofactor Requirement of the Microsomal Side-Chain Hydroxylation System Acting on 5β-Cholestane-3α,7α,12α,25-tetrol[a]

	Rate of formation of 5β-cholestanepentols, pmol/mg protein/10 min
Complete system[b]	13,500
Minus NADPH	446
Minus NADPH plus NADH	2,250
Under N$_2$	1,275
Boiled microsomal fraction	20

[a] Rat liver microsomes.
[b] Standard assay system (43).

TABLE XVIII. Subcellular Distribution of Hepatic Enzyme Activity Transforming 5β-Cholestane-3α,7α,12α,24,25-pentols to Cholic Acid[a]

| | Human, pmol/mg protein/min | | | | Rat, pmol/mg protein/min | |
| | Normal | | CTX | | | |
	24α-pentol[b]	24β-pentol[c]	24α-pentol[b]	24β-pentol[c]	24α-pentol[b]	24β-pentol[c]
Whole homogenate	13.3	430	9.1	303	15.0	450
Mitochondria[d]	1.9	73.1	1.3	42.0	2.2	72.6
Microsomes	0.8	20.7	0.5	11.9	1.2	20.6
100,000g supernatant	25.1	916	17.2	731	27.2	1084
Mitochondria + 100,000g supernatant	5.6	203	4.3	182.7	5.4	216
Microsomes + 100,000g supernatant	10.5	382	6.8	282	11.0	452

[a] Subcellular fractionation was performed as described (43).
[b] 5β-Cholestane-3α,7α,12α,24α,25-pentol.
[c] 5β-Cholestane-3α,7α,12α,24β,25-pentol.
[d] Partially broken mitochondria.

TABLE XIX. Substrate Specificity of Soluble Enzyme System Transforming
5β-Cholestanepentols into Cholic Acid[a]

Substrate	Rate of cholic acid production, pmol/mg protein/min		
	Human liver		Rat liver
	Control	CTX	
5β-Cholestane-3α,7α,12α,23ξ,25-pentol[b]	33.8	40.3	37.5
5β-Cholestane-3α,7α,12α,24α,25-pentol[b]	25.1	17.2	22.8
5β-Cholestane-3α,7α,12α,24β,25-pentol	916	731	983
5β-Cholestane-3α,7α,12α,25,26-pentol[b]	1.9	1.5	2.6

[a] Standard assay system (43).
[b] The small amounts of radioactivity found in the cholic acid area of the TLC plate following incubation with 5β-cholestane-3α-7α,12α,23ξ,25-pentol, 5β-cholestane-3α,7α,12α,24α,25-pentol, and 5β-cholestane-3α,7α,12α,25,26-pentol were not positively identified as cholic acid.

been done it may be speculated that the desmolase which catalyzes cleavage of the side chain between carbons 24 and 25 is a dehydrogenase/hydroxylase as has been postulated for the mechanism of adrenal hormone synthesis.

The results of this *in vitro* study suggest that there exists an alternate pathway of cholic acid biosynthesis in man and in the rat involving the 25-hydroxylation of 5β-cholestane-3α,7α,12α-triol (V, Fig. 10). This pathway proceeds via 5β-cholestane-3α,7α,12α,25-tetrol (VIII) and 5β-cholestane-3α,7α,12α,24β,25-pentol (XI) and does not involve 5β-cholestanoic acids as intermediates.

TABLE XX. Cofactor Requirement of Soluble Enzyme Activity Transforming
5β-Cholestane-3α,7α,12α,24β,25-pentol into Cholic Acid

Cofactors added	Rate of cholic acid formation pmol/mg protein/min		
	Human liver		Rat liver
	Control	CTX	
Complete system[a]	1082	811	995
Minus NAD+	30.7	21.3	24.1
Minus NAD+ plus NADP+ (3 μmoles)	165.9	121.8	110.6
Boiled 100,000g supernatant	21.6	14.7	20.0

[a] Standard assay system (43).

In summary, patients with cerebrotendinous xanthomatosis suffer from progressive neurological impairment, premature atherosclerosis, tendon and tuberous xanthomas, cataracts, and possibly endocrine insufficiency because of the excessive deposition of cholesterol and its 5α-dihydro derivative, cholestanol, in most tissues. The accumulation of cholesterol and cholestanol results from the overproduction of both sterols by the liver. The disease is inherited as a recessive disorder and appears to be related to defective cholesterol side chain oxidation. As a consequence, 25-hydroxylated bile alcohols are excreted and total bile acid production is low with a decline in the chenodeoxycholic acid content of bile. The bile alcohols excreted by CTX subjects have been identified as 5β-cholestane-$3\alpha,7\alpha,12\alpha$, 25-tetrol, 5β-cholestane-$3\alpha,7\alpha,12\alpha,23\xi,25$-pentol, and 5β-cholestane-3α, $7\alpha,12\alpha,24\alpha,25$-pentol. 5β-Cholestane-$3\alpha,7\alpha,12\alpha,25$-tetrol is converted to cholic acid in man *in vivo* and *in vitro,* thus suggesting a new pathway for side chain oxidation that involves C-25 hydroxy intermediates. Finally, treatment of CTX subjects with chenodeoxycholic acid inhibits cholesterol and cholestanol production and reduces plasma cholestanol levels (38). This treatment may prevent further progression of this disease and perhaps result in the reversal of the clinically significant lesions.

REFERENCES

1. W. R. Harlan, J. B. Graham, and E. H. Estes, *Medicine* **45**, 77 (1966).
2. J. Piper, and L. Orrild, *Am. J. Med.* **21**, 34 (1956).
3. D. S. Fredrickson, R. I. Levy, and R. S. Lees, *N. Engl. J. Med.* **276**, 34, 94, 148, 215, 273 (1967).
4. J. D. Wilson, *Circ. Res.* **12**, 472 (1963).
5. A. Bhattacharyya and W. E. Connor, *J. Clin. Invest.* **53**, 1033 (1974).
6. L. van Bogaert, H. J. Scherer, and E. Epstein, Paris, Masson Cie. (1937).
7. G. Salen, *Ann. Intern. Med.* **75**, 843 (1971).
8. C. Schneider, *Allg. Z. Psychiatr.* **104**, 144 (1936).
9. E. Epstein and K. Lorenz, *Klin. Wochenschr.* **16**, 1320 (1937).
10. G. Guillain, I. Bertrand, and M. Godet-Guillain, *Rev. Neurol.* (Paris) **74**, 249 (1942).
11. D. Vindetti, *Chir. Organi. Mov.* **34**, 429 (1950).
12. A. Giampalmo, *Acta Neurol. Belg.* **54**, 786 (1954).
13. W. Stein, and S. Czuczwar, *Neurol. Neurochir. Pol.* **9**, 599 (1959).
14. J. C. Ortiz de Zarate, *Rev. Neurol.* **19**, 159 (1961).
15. M. Philippart and L. van Bogaert, *Arch. Neurol.* **21**, 603 (1969).
16. B. M. Derby, H. Pogocar, C. Mueckenhausen, H. W. Moser, and E. P. Richardson, Jr., *J. Neuropathol. Exp. Neurol.* **29**, 139 (1970).
17. J. R. Schimschock, E. C. Alvord Jr., and P. D. Swanson, *Arch. Neurol.* **19**, 688 (1968).
18. J. D. Hughes and T. W. Meriwether, *South. Med. J.* **64**, 311 (1971).
19. W. R. Harlan Jr. and W. J. S. Still, *N. Engl. J. Med.* **278**, 416 (1968).

20. H. Farpour and M. Mahloudji, *Arch. Neurol.* **32**, 233 (1975).
21. A. Schreiner, C. Hoper, and S. Skrede, *Acta Neurol. Scand.* **51**, 405 (1975).
22. J. H. Menkes, J. R. Schimschock, and P. D. Swanson, *Arch. Neurol.* **19**, 47 (1968).
23. T. A. Miettinen, E. H. Ahrens, Jr., and S. M. Grundy, *J. Lipid Res.* **6**, 397 (1965).
24. S. M. Grundy, E. H. Ahrens, Jr., and T. A. Miettinen, *J. Lipid Res.* **6**, 411 (1965).
25. DeW. Goodman and R. P. Noble, *J. Clin. Invest.* **47**, 231 (1968).
26. E. Gurpide, J. Mann, and E. Sandberg, *Biochemistry* **3**, 1250 (1964).
27. G. Salen and S. M. Grundy, *J. Clin. Invest.* **52**, 2822 (1973).
28. T. Setoguchi, G. Salen, G. S. Tint, and E. H. Mosbach. *J. Clin. Invest.* **53**, 1393 (1974).
29. P. Samuel, and S. Lieberman, *J. Lipid Res.* **14**, 189 (1973).
30. DeW. Goodman, R. P. Noble, and R. B. Dell, *J. Lipid Res.* **14**, 178 (1973).
31. G. Nicolau, S. Shefer, G. Salen, and E. H. Mosbach, *J. Lipid Res.* **15**, 94 (1974).
32. G. Salen, and A. Polito, *J. Clin. Invest.* **51**, 134 (1972).
33. J. W. Cornforth, R. H. Cornforth, C. Donninger, and G. Popják, *Proc. R. Soc. London, Ser. B* **163**, 492 (1965–1966).
34. R. S. Rosenfeld, B. Zumoff, and L. Hellman, *J. Lipid Res.* **8**, 16 (1967).
35. S. Shefer, S. Hauser, and E. H. Mosbach, *J. Lipid Res.* **7**, 763 (1966).
36. I. Björkhem, and K. E. Karlmar, *Biochim. Biophys. Acta* **337**, 129 (1974).
37. G. Salen, *Clin. Res.* **22**, 367A (1974).
38. G. Salen, T. W. Meriwether, and G. Nicolau, *Biochem. Med.* **14**, 57 (1975).
39. G. Nicolau, S. Shefer, G. Salen, and E. H. Mosbach, *J. Lipid Res.* **15**, 146 (1974).
40. G. Salen, S. Shefer, T. Setoguchi, and E. H. Mosbach. *J. Clin. Invest.* **56**, 226 (1975).
41. T. Chronholm and G. Johansson, *Eur. J. Biochem.* **16**, 373 (1970).
42. I. Björkhem, J. Gustafsson, G. Johansson, and B. Persson, *J. Clin. Invest.* **55**, 478 (1975).
43. S. Shefer, F. W. Chen, B. Dayal, S. Hauser, G. S. Tint, G. Salen, and E. H. Mosbach. *J. Clin. Invest.* **57**, 897 (1976).
44. T. Masui and E. Staple, *J. Biol. Chem.* **241**, 3889 (1965).
45. J. Gustafsson, *J. Biol. Chem.* **250**, 8243 (1975).
46. S. Shefer, B. Dayal, G. S. Tint, G. Salen, and E. H. Mosbach, *J. Lipid Res.* **16**, 280 (1975).

Chapter 7

BILE ACIDS AND INTESTINAL CANCER*

Norman D. Nigro and Robert L. Campbell

Department of Surgery
Wayne State University School of Medicine
Detroit, Michigan

I. INTRODUCTION*

There is evidence that bile acids may play a significant role in the etiology of cancer of the large bowel. Epidemiological studies indicate that the disease is due largely to environmental factors, and recent experimental studies tend to support this concept. The studies show that certain dietary habits affect bile acid metabolism in a way which may have significance in intestinal carcinogenesis. It has been known for some time that bile acids are potentially carcinogenic due to their steric similarity to such carcinogenic hydrocarbons as 3-methylcholanthrene (1,2) and the possible conversion of deoxycholic acid† to the same compound by gut microflora (3–5). The purpose of this chapter is to review the evidence which tends to implicate bile acids in intestinal carcinogenesis.

* Part of the investigations reported in this chapter are supported by the Matilda R. Wilson Fund, Detroit, Michigan.

† The following systematic names are given bile acids referred to by trivial names: chenodeoxycholic acid, $3\alpha,7\alpha$-dihydroxy-5β-cholanoic acid; cholic acid, $3\alpha,7\alpha,12\alpha$-trihydroxy-5β-cholanoic acid; deoxycholic acid, $3\alpha,12\alpha$-dihydroxy-5β-cholanoic acid; hyodeoxycholic acid, $3\alpha,6\alpha$-dihydroxy-5β-cholanoic acid; 12-ketolithocholic acid, 3α-hydroxy-12-keto-5β-cholanoic acid; lithocholic acid, 3α-hydroxy-5β-cholanoic acid; β-muricholic acid, $3\alpha,6\beta,7\beta$-trihydroxy-5β-cholanoic acid; taurocholic acid, $3\alpha,7\alpha,12\alpha$-trihydroxy-5β-cholan-24-oyltaurine; taurodeoxycholic acid, $3\alpha,12\alpha$-dihydroxy-5β-cholan-24-oyltaurine.

II. EPIDEMIOLOGICAL OBSERVATIONS

Epidemiological studies suggest that environment plays the leading role in the development of cancer of the large bowel. These surveys show that the incidence of large bowel cancer differs widely around the world. The disease is common in the United States, Canada, Britain, and northwest Europe, but uncommon in Asia, Africa, and parts of South America (6). The incidence rate is higher in the northeastern part of the United States than in the South, and it is more common in urban than in rural areas (7).

Epidemiological studies of migratory populations show that people who move from a low to a high incidence area soon develop the increased risk of the host country. Examples of such studies involve the migration of people from Poland to the United States (8), Japanese to Hawaii and California (9), and Blacks from southern rural regions to northern urban areas (10).

It is reasonable to assume that the environmental factor in cancer of the large bowel is diet. Studies of dietary habits show that in the high incidence areas, the diet is low in residue and high in animal fat and protein, especially beef. While in the low incidence areas, it is high in residue and low in animal fat. The correlation of increased incidence with beef consumption is particularly striking (10).

There is no doubt that cancer in general develops through an interplay of both genetic and environmental factors. The weight of the two elements varies in individuals and this is true in cancer of the large bowel. For instance, in familial polyposis all patients having this syndrome develop cancer of the large bowel unless this portion of the intestinal tract is removed prophylactically. Here the genetic factor is clearly dominant. The genetic element also may be fairly significant in families who have a high incidence of cancer in general. But the number of such people is quite small, and epidemiological studies, especially the migratory surveys cited above, show that in cancer of the large bowel the dietary factor is dominant in the vast majority of people. This is, indeed, fortunate because the identification of an important dietary causative element might well form the basis for preventive measures resulting in a significant reduction in the incidence of this common and lethal form of cancer.

III. FECAL BILE ACIDS AND COLON CANCER IN THE HUMAN

At the beginning of the present decade, new interest developed in the possible cancer-producing properties of bile acids. This was generated by chemical evidence relating fecal acidic steroid composition to the high incidence of cancer of the colon, an organ in which bile acids are in intimate contact. Hill and his associates (11,12) were first to suggest that the epidemiological correlation between diet and colon cancer might be explained by the involvement of bile steroids. They analyzed the fecal material of people from areas of high and low colon cancer incidence. A correlation was found between a high cancer incidence and a high concentration of fecal acid steroids. Further, there were more anaerobic microbes capable of dehydroxylating cholic acid in the 7α-position to form deoxycholic acid. In a further report, Hill (13) found that both the fecal bile acid and neutral steroid concentration in adults was dependent on the amount of dietary fat. As the total concentration of fecal bile acids increases, so does the proportion of secondary bile acids formed by the metabolic activity of the intestinal bacteria. These results support the working hypothesis (14,15) that diet influences the bacterial flora of the gut as well as the types and amounts of compounds with which this flora is in contact. In turn, the interaction of the flora with dietary compounds and digestive secretions, in particular bile steroids, may produce carcinogens or cocarcinogens which act upon colonic tissue.

Additional evidence in support of this hypothesis has been presented by Reddy and Wynder (16,17). The daily excretion of bile acids and the degree of microbial activity in the gut as measured by β-glucuronidase activity in the feces was determined in a number of populations. Americans consuming a diet containing large amounts of fat and animal protein excreted 4- to 5-fold more lithocholic acid and 2.5- to 4-fold more deoxycholic acid than did Japanese, Chinese, and American vegetarians. The diet of these latter groups is characterized by low amounts of fat and animal protein. The β-glucuronidase activity of the feces was significantly greater in Americans eating a typical Westernized diet. Of the groups studied, these Americans had the highest incidence of colon cancer.

In early 1975 Hill et al. (18) reported the results of a further study of fecal bile acids and anaerobic bacteria in patients with colonic cancer. The results indicated that a larger proportion (82%) of colon cancer patients had high levels of fecal bile acids (>6mg/g fecal dry weight) than did patients with other diseases (17%). Furthermore, bacteriological studies showed that in 70% of colon cancer patients, a higher number of fecal clostridia able to dehydrogenate the bile acid nucleus was associated with high fecal bile acid levels. This association was present in only 9% of control patients. Detailed analysis indicated that significantly more degraded bile acids were present in the feces of the colon cancer patients than in the control groups. The investigators suggested that large amounts of bile acid derivatives produced by certain anaerobic microbes were the causative agents in colon cancer.

Reddy et al. (49) summarized studies related to the metabolic epidemiology of large bowel cancer. They concluded that a strong association has been established between microbially modified bile acids and cholesterol metabolites and the risk of colon cancer among different populations. As yet, however, this association has not been shown to be causative in nature.

In another report, Reddy *et al.* (50) studied the effects of high-risk and low-risk diets for colon carcinogenesis on fecal microflora and steroids in man. The results showed that the high-animal-protein and -fat diets consumed by high-risk populations for colon cancer affect the composition of intestinal microflora as well as the level of certain bile acids that may in some way influence carcinogenesis in the colon.

IV. INVESTIGATIONS OF INTESTINAL CARCINOGENESIS IN ANIMALS

A. Development of Animal Models

Spontaneous cancers of the intestinal tract are rare in laboratory animals. Furthermore, their induction by experimental means with a high degree of yield and specificity has not been possible until recently.

In 1963, Laqueur and associates (19), investigating the toxicology of cycasin (β-D-glucosyloxyazoxymethane), a natural product of the cycad plants, found it to be carcinogenic when fed to rats for a prolonged period. This compound induced tumors in the liver, kidney, and intestine. Subsequently, they found this carcinogenic effect dependent upon the bacterial flora of the intestinal tract (20). After the chemical structure of cycasin was established, Matsumoto *et al.* (21) synthesized its aglycone, methylazoxymethanol (MAM). It was found that MAM induced tumors independent of the route of administration or of bacteria. It was concluded that MAM was the proximate carcinogen of cycasin (22).

Following the discovery of the carcinogenicity of cycasin and of MAM, Druckrey (23) tested the carcinogenic properties of two closely related compounds, 1,2-dimethylhydrazine and azoxymethane. They found these compounds to be effective and highly selective carcinogens for the intestinal tract of rodents especially the large intestine. When given subcutaneously at weekly intervals they uniformly induced intestinal tumors in 184–380 days depending on dosage. The tumors were similar, both grossly and histologically, to human large bowel cancers.

Dimethylhydrazine, azoxymethane, and methylazoxymethanol have thus been established as useful intestinal carcinogens for rats, mice, and hamsters. They are effective when given subcutaneously, orally, or rectally, though the subcutaneous route is best. The model most often used at present is one developed by the subcutaneous injection of dimethylhydrazine into mice or rats.

Cancer of the large bowel has recently been induced with a high degree of success by the rectal instillation of N-methyl-N'-nitro-N-nitrosoguanidine (24) or by N-methylnitrosourea (25) in mice. These chemicals appear to have a direct action on the intestinal mucosa and therefore, provide useful additional models for studies of the intestinal carcinogenic process in the experimental animal.

B. Studies on Bile Acid Involvement

1. Drugs

Observations suggesting that alterations in bile acid metabolism by drugs affect intestinal carcinogenesis in the experimental animal have been reported by Nigro and associates (26). Rats were fed a diet containing 2% cholestyramine, a nonabsorbable anion-exchange resin, known to increase the amount of fecal bile acids (27). The animals were given one of the following carcinogens subcutaneously at weekly intervals over a nine month period; 1,2-dimethylhydrazine, azoxymethane, or methylazoxymethanol.

A striking change in the number and distribution of intestinal tumors was observed in the rats fed cholestyramine with each of the three carcinogens. With azoxymethane, for example, there were 6.8 tumors per rat on the normal diet, while rats fed cholestyramine developed 13.5 tumors per rat. All were malignant. As can be seen in Fig. 1, the increase in the number of cancers occurred in the lower segments of the intestinal tract, especially in the distal half of the large intestine. The effect of cholestyramine was the same with the other two carcinogens.

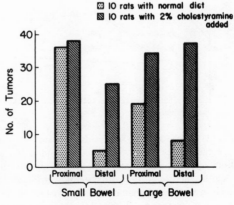

Fig. 1. Distribution of intestinal tumors in rats given azoxymethane and fed a normal diet or one containing 2% cholestyramine. Rats given the cholestyramine developed twice as many tumors; the greatest increase was in tumors in the large intestine. [From Nigro *et al.* (26), reprinted with permission of J. B. Lippincott Company.]

Nigro *et al.* (28) have also studied the effect of 2% dietary cholestyramine on the total bile acid content of tissues from various regions of the intestine and in the feces of rats given azoxymethane. The carcinogen and cholestyramine both increased the total bile acid content of the samples. The two agents together, however, did not give an additive effect. The greatest increase over controls in total bile acid level was observed in the feces and distal colonic tissues of animals given either azoxymethane or cholestyramine and azoxymethane together (Fig. 2). Carey and Williams (29) found that increased bile acid excretion in the human caused by cholestyramine was the result of a sharp increase in deoxycholic acid content of the feces. Huff *et al.* (30) reported a similar finding in the rat. The available data suggest that an alteration in the amount and types of bile acids in the large bowel may be the important factor in cholestyramine's carcinogenic enhancing effect.

Asano *et al.* (51) studied the effect of dietary cholestyramine on 1,2-dimethylhydrazine-induced intestinal cancer in germ-free Sprague-

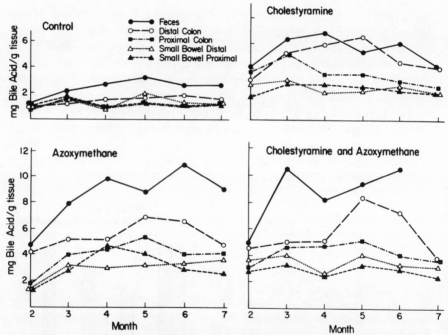

Fig. 2. Levels of total bile acids in tissues from various segments of intestine or feces of Sprague-Dawley rats given 2% cholestyramine and/or azoxymethane. Bile acids were extracted from aqueous homogenates by a method modified from Iwata and Yamasaki (32) and quantitated by the use of 3α-hydroxysteroid dehydrogenase (33). The highest concentration of bile acids in the treated animals was observed in the distal colon and feces. These levels were significantly (average $p < 0.0005$) different than those of the control.

TABLE I. Total Fecal Bile Acids from Normal Rats Compared to Rats with a Surgically Implanted Bile Duct[a]

Week	Fecal material, mg/g (mean + S.E.)		p[c]
	Normal rats (4)[b]	Rats with implanted ducts (5)[b]	
1	1.27 ± 0.22	2.97 ± 0.54	< 0.05
2	1.09 ± 0.21	2.60 ± 0.47	< 0.025
3	1.08 ± 0.10	2.23 ± 0.13	< 0.0005
4	1.61 ± 0.08	2.76 ± 0.19	< 0.005
5	1.10 ± 0.17	2.15 ± 0.12	< 0.005

[a] The total bile acid content of the fecal material from normal and bile duct implanted animals over a five week period. The bile acids were extracted from aqueous homogenates by a method modified from Iwata and Yamasaki (32) and quantitated by the use of 3α-hydroxysteroid dehydrogenase (33). The rats with implanted bile ducts had significantly greater concentrations of total fecal bile acids [data from Chomchai et al. (31)].
[b] Number of animals used per week.
[c] Significance level as determined by Student's t test (34).

Dawley rats. Tumor incidence increased from 9.3 tumors per rat to 16.5 tumors per rat when cholestyramine was administered. The authors concluded that some aspect of the dynamic status of bile acid metabolism has a direct relation to intestinal carcinogenesis and that the importance of secondary bile acids in colon carcinogenesis may be overestimated.

2. Mechanical

The content of bile acids in the lower intestinal tract of rats was altered by mechanical means in a study by Chomchai et al. (31). The distal end of the bile duct was divided close to its entrance into the duodenum. The duct was then anastomosed to the midpart of the small intestine. Therefore, bile in the operated animals entered the intestine at a much lower level than normal, bypassing the entire proximal half of the small intestine. The total bile acid concentration in feces from these animals was found to be twice normal (Table I). Azoxymethane was given subcutaneously to a group of rats operated upon in this manner. They developed significantly more tumors than the unoperated control animals (Fig. 3). In addition, the enhanced tumor formation occurred mainly in the large intestine. The effect was the same as that obtained with cholestyramine, supporting the concept that the drug's effect on carcinogenesis is due to its action on bile acids.

3. Dietary

The addition of fat to the diet of rats given an intestinal carcinogen has been found to enhance tumor formation and to alter the composition of the

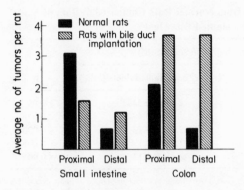

Fig. 3. Distribution of azoxymethane-induced intestinal tumors in normal and bile duct implanted rats.

fecal steroids excreted. Reddy *et al.* (35) have given rats semipurified diets with 0.5% (low-fat), 4% (normal-fat), or 20% (high-fat) corn oil and used 1,2-dimethylhydrazine as the carcinogen. Ninety percent of the animals on the high fat diet, but only about sixty percent of the low-fat or normal-fat diet animals developed colon tumors. The average number of tumors per animal with the high-fat diet was significantly increased over that of the animals fed lesser amounts of fat. The rats fed a high-fat diet excreted significantly more deoxycholic and β-muricholic acids per kilogram body weight in 24 h than did animals fed the other diets. It appears the corn oil diet altered the microbial activity of the gut causing an increase in bile acid degradation. These products might then be acting as promoters to carcinogenesis.

Working on the same hypothesis, Nigro *et al.* (36) fed rats on azoxymethane a diet containing 35% beef fat. These animals developed significantly more intestinal tumors than the controls (Fig. 4). Furthermore, the tumors in the high-fat diet group were larger and more malignant than those in the control animals.

The beef fat caused significant changes in the composition of the major bile acids found in the feces (Table II). After two months of treatment with azoxymethane, the rats on the high-fat diet had significantly greater amounts of microbially degraded fecal bile acids, i.e., deoxycholic and 12-ketolithocholic acids, than did normal diet animals. These degraded bile acids may be promoting the carcinogenic activity of azoxymethane. The results of this study give a possible biochemical explanation for the epidemiological correlation between beef consumption and a high incidence of large bowel cancer.

Other dietary factors may also have relevance to the etiology of colon cancer. Fiber is one of these: Epidemiological studies show important differences in the amount of fiber present in the diets of people in various sections of the world. Some studies appear to indicate that fiber intake may be of some importance in intestinal carcinogenesis (38,39). However, other observations show little correlation between fiber intake and the incidence of colon cancer (40).

There is experimental evidence that dietary fiber may have an important effect on bile acids. The *in vitro* binding of radioactively labeled sodium taurocholate and glycocholate by bran and cellulose, as well as several other natural and synthetic fibers, has been studied by Kritchevsky and Story (41). They found that natural fibers such as bran, alfalfa, etc., had a greater capacity to bind bile salts than did the synthetic fibers.

In addition, Ward *et al.* (42) tested the effect of dietary cellulose on azoxymethane-induced intestinal cancers in rats. They found fewer small-bowel tumors in the animals on the high cellulose diet compared to those on lower residue diets. This effect did not occur in the large intestine. Therefore, it appears from this study that the effect of the fiber was similar to that of bile acid sequestering agents such as cholestyramine. There is a need for further studies to clarify the relationship of dietary fiber to colon carcinogenesis, particularly the role of fiber in altering bile metabolism and the colonic bacterial flora.

Nigro *et al.* (52) studied the effect of a high-beef-fat diet on the composi-

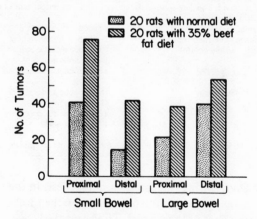

Fig. 4. Distribution of intestinal tumors in rats treated with azoxymethane and fed a normal- or high-beef-fat diet. The average number of tumors in the entire small intestine in the normal diet group was 2.80 ± 1.01, whereas it was 5.95 ± 1.03 ($p < 0.025$) in the high-fat diet group. In the entire large intestine the average tumor number was 3.10 ± 0.58 for the normal diet animals and 4.55 ± 0.46 ($p < 0.05$) for the high-fat group. [From N. D. Nigro *et al.* (36).]

TABLE II. Composition of Fecal Bile Acids Excreted by Rats Fed Normal or Beef Fat Diets and Given Azoxymethane[a]

	Controls, mg/g dry feces; mean ± S.E.		Azoxymethane treated, mg/g dry feces; mean ± S.E.	
Bile acid	Normal diet (5)[b]	Fat diet (5)[b]	Normal diet (5)[b]	Fat diet (5)[b]
Chenodeoxycholic	0.040 ± 0.012	0.108 ± 0.030	0.037 ± 0.008	0.152 ± 0.023
Cholic	0.078 ± 0.020	0.063 ± 0.010	0.127 ± 0.015	0.095 ± 0.017
Deoxycholic	0.300 ± 0.051	0.340 ± 0.040	0.289 ± 0.036	0.739 ± 0.139
Hyodeoxycholic	0.251 ± 0.055	0.176 ± 0.017	0.260 ± 0.037	0.311 ± 0.025
12-Ketolithocholic	0.186 ± 0.024	0.161 ± 0.028	0.239 ± 0.037	0.346 ± 0.042
Lithocholic	0.132 ± 0.021	0.137 ± 0.015	0.141 ± 0.019	0.168 ± 0.120
Total	0.986 ± 0.152	0.993 ± 0.124	1.093 ± 0.119	1.810 ± 0.210

[a] Fecal samples were collected from pairs of animals in each group over a 24 h period. At the time of the collection, the animals had been fed their respective diets (normal = Purina Rat chow; fat = granular Purina Rat chow with 35% beef fat by weight) for two months. Those animals receiving azoxymethane were given weekly subcutaneous injections (8 mg/kg body weight) over the two month period. The bile acids were extracted from the feces by the method of Grundy et al. (37) with slight modifications. [24-^{14}C] Cholic acid was used as internal standard to determine extraction recoveries. The fecal bile acids were methylated before thin layer chromatography by overnight reaction at room temperature with 14% (w/v) BF_3–MeOH. The trimethylsilyl (TMS) ethers of the methyl esters were prepared using Bis(trimethylsilyl) trifluoroacetamide. The TMS derivatives were analyzed with a Varian Aerograph model 2700 gas chromatograph on a glass column packed with 3% QF-1 on 100–120 mesh Gas-Chrom Q. Column temperature was 240°C with the flame ionization detector at 275°C. Quantitation was achieved with 5α-cholestane as internal standard. The data were analyzed statistically by the Student's t test (34).
Significant differences observed:

Fat control vs. normal control	Chenodeoxycholic acid	$p < 0.05$
Normal azoxymethane vs. normal control	No significant differences	
Fat azoxymethane vs. normal control	Chenodeoxycholic acid	$p < 0.005$
	Deoxycholic acid	$p < 0.025$
	12-Ketolithocholic acid	$p < 0.01$
	Total	$p < 0.025$
Fat azoxymethane vs. fat control	Deoxycholic acid	$p < 0.025$
	Hyodeoxycholic acid	$p < 0.005$
	12-Ketolithocholic acid	$p < 0.01$
	Lithocholic acid	$p < 0.025$
	Total	$p < 0.01$

[b] Number of pooled fecal samples studied, each sample the combination of two animals.

tion of fecal bile acids during intestinal carcinogenesis in the rat. The results showed that there was not a significant increase in the total concentration of fecal bile acids as a result of the fat diet. However, the level of microbial degradation of cholic acid to deoxycholic acid was increased by the fat diet. Animals fed this diet and given azoxymethane were previously found to produce more tumors than control diet animals.

4. Fecal Stream

It is reasonable to expect that the fecal stream is an important factor in carcinogenesis of the intestine. Its obligatory nature, however, has been a subject of dispute. In Wistar rats with the fecal stream of the colon diverted by a colostomy, Navarette and Spjut (43) found that subcutaneous injections of 3,2′-dimethyl-4-aminobiphenyl did not produce tumors beyond the colostomy site. Similar results have been reported by Cleveland *et al.* (44) in rats having defunctionalized colon segments. These results suggested that the fecal stream is essential to tumorigenesis. However, Wittig *et al.* (45) reported the formation of cancer in the colon distal to colostomies when 1,2-dimethylhydrazine was given as the carcinogen. These contrasting results suggest that intestinal carcinogens may have differing routes of action or their activity is affected differently by the fecal stream.

To study the latter possibility, Campbell *et al.* (46) have investigated the importance of the fecal stream on the induction of colon tumors by azoxymethane. The removal of the fecal stream from a part of the colon by colostomy significantly decreased the average number of tumors per rat in that segment of colon to 2.3 ± 0.6. This is in contrast to 4.9 ± 0.6 tumors per rat in the same length of colon in the unoperated controls. Two percent dietary cholestyramine tended to increase the colonic tumor yield in the intact animal but did not affect the number of tumors in the colons of the animals with colostomies (Fig. 5). The absence of luminal bile acids in the defunctionalized colon of colostomy animals was suggested as a reason for the change in tumor incidence. Likewise, the lack of a fecal stream, through which cholestyramine manifests its effect on bile acids, prevented this drug from enhancing carcinogenesis in the defunctionalized colon.

Similar studies have been conducted in germ-free animals. Reddy *et al.* (47) found more tumors in conventional than germfree rats given 1,2-dimethylhydrazine. These investigators suggested that the difference in tumor susceptibility may be explained by the promoting activity of certain unconjugated bile acids which are absent in the fecal stream of the germ-

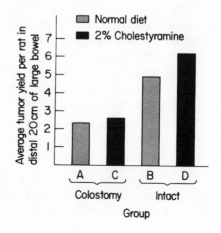

Fig. 5. Average tumor yield per rat in the distal 20 cm of large bowel for normal rats (functional colon) and rats with colostomies (nonfunctional colon). The differences observed in the tumor yield between Groups A and B and also between Groups C and D are significant ($p < 0.005$ and $p < 0.0005$, respectively). [From R. L. Campbell *et al.* (46) by permission of Cancer Research, Inc.]

free animal. There is further direct experimental evidence supporting the hypothesis that certain bile acids. or their metabolites, present in the fecal stream may be promoters of intestinal carcinogenesis. After a single rectal dose of N-methyl-N'-nitro-N-nitrosoguanidine, Narisawa *et al.* (48) have found that multiple rectal instillations of lithocholic acid and taurodeoxycholic acid in peanut oil increased the incidence of colon tumors twice over that with the carcinogen alone. The two bile acids instilled without the carcinogen produced no tumors. In this animal model, the bile acids, or their metabolites, produced by microflora appear to be acting as promoting agents to carcinogenesis but are not carcinogenic to the colonic mucosa themselves.

V. SUMMARY

The geographic pattern of the incidence of cancer of the large bowel suggests that an environmental factor, probably diet, is the important causative agent in this form of cancer. This epidemiological clue which implicates a high consumption of fat is supported by experimental studies in humans.

Animal studies also support the concept that the ingestion of large amounts of fat, especially of animal origin, is an important element in the causation of intestinal cancer. When large amounts of fat, particularly beef, are fed to rats on an intestinal carcinogen, tumor formation is accelerated. Studies provide considerable evidence that this effect is mediated through a mechanism involving the consequences of increased bile acid secretion and degradation.

Clarification of the entire process by which intestinal cancer develops, including an understanding of all genetic, biochemical, and environmental factors, will take years of investigation. On the other hand, continued study of the role bile acids play in intestinal carcinogenesis may well provide evidence which would allow the development of preventative measures to lower the cancer incidence. The prospects for this more immediate, limited objective are good, and indeed more promising than advances in the field of earlier diagnosis or improved treatment.

REFERENCES

1. H. Weiland and E. Dane, *Hoppe-Seyler's Physiol. Chem. Z.* **219**, 240 (1933).
2. J. W. Cook and G. A. D. Haslewood, *Chem. and Ind.* (Rev.) **11**, 758 (1933).

3. V. C. Aries, J. S. Crowther, B. S. Drasar, M. J. Hill, and R. E. O. Williams, *Gut* **10**, 334 (1969).

4. P. Goddard and M. J. Hill, *Biochim. Biophys. Acta* **280**, 336 (1972).

5. P. Goddard and M. J. Hill, *Trans. Biochem. Soc.* **1**, 1113 (1973).

6. R. Doll, *Br. J. Cancer* **23**, 1 (1969).

7. B. Jansson, G. B. Siebert, J. F. Speer, *Cancer* **36**, 2373 (1975).

8. J. Staszewski, *Recent Results Cancer Res.* **39**, 85 (1972).

9. W. Haenszel, J. W. Berg, M. Segi, M. Kurihara, and F. Locke, *J. Nat. Cancer Inst.* **51**, 1765 (1973).

10. J. W. Berg, M. Howell, and S. Silverman, *Health Serv. Rep.* **88**, 915 (1973).

11. M. J. Hill, B. S. Drasar, V. Aries, J. S. Crowther, G. Hawksworth, and R. E. O. Williams, *Lancet* **1**, 95 (1971).

12. M. J. Hill and V. C. Aries, *J. Pathol.* **104**, 129 (1971).

13. M. J. Hill, *J. Pathol.* **104**, 239 (1971).

14. B. S. Drasar and M. J. Hill, *Am. J. Clin. Nutr.* **25**, 1399 (1972).

15. V. C. Aries, J. S. Crowther, B. S. Drasar, M. J. Hill, and R. E. O. Williams, *Gut* **10**, 334 (1969).

16. B. S. Reddy and E. L. Wynder, *J. Nat. Cancer Inst.* **50**, 1437 (1973).

17. E. L. Wynder and B. S. Reddy, *Cancer* **34**, 801 (1974).

18. M. J. Hill, B. S. Drasar, R. E. O. Williams, T. W. Meade, A. G. Cox, J. E. P. Simpson, and B. C. Morson, *Lancet,* 535, (1975).

19. G. L. Laqueur, O. Mickelson, M. G. Whiting, and L. Kurland, *J. Nat. Cancer Inst.* **31**, 919 (1963).

20. G. L. Laqueur, *Fed. Proc.* **23**, 1386 (1964).

21. H. Matsumoto, T. Nagahama, and H. O. Larson, *Biochem. J.* **95**, 13c (1965).

22. G. L. Laqueur, E. G. McDaniel, and H. Matsumoto, *J. Nat. Cancer Inst.* **39**, 355 (1967).

23. H. Druckrey, *in* "Carcinoma of the Colon and Antecedent Epithelium" (W. J. Burdette, ed.), p. 267, Charles C. Thomas, Springfield (1970).

24. J. H. Weisburger, *Dis. Colon Rectum* **16**, 431 (1973).

25. T. Narisawa and J. H. Weisburger, *Proc. Soc. Exp. Biol. Med.* **148**, 166 (1975).

26. N. D. Nigro, N. Bhadrachari, and C. Chomchai. *Dis. Colon Rectum* **16**, 438 (1973).

27. D. Kritchevsky, *in* "The Bile Acids: Chemistry, Physiology, and Metabolism" (P. P. Nair and D. Kritchevsky, eds.), Vol. 2, p. 273, Plenum Press, New York (1973).

28. N. D. Nigro, V. M. Sardesai, C. Chomchai, R. L. Campbell, and H. S. Provido, *Fed. Proc.* **33**, 260 (1974).

29. J. B. Carey, Jr. and G. Williams, *J. Am. Med. Assoc.* **176**, 432 (1961).

30. J. W. Huff, J. L. Gilfillan, and V. M. Hunt, *Proc. Soc. Exp. Biol. Med.* **114**, 352 (1963).

31. C. Chomchai, N. Bhadrachari, and N. D. Nigro, *Dis. Colon Rectum* **17**, 310 (1974).

32. T. Iwata and K. Yamasaki, *J. Biochem.* (Tokyo) **56**, 424 (1964).

33. P. Talalay, *in* "Methods in Enzymology" (S. P. Colowick and N. O. Kaplan, eds.), Vol. V, p. 512, Academic Press, Inc., New York (1962).

34. W. Beyer, Handbook of Tables for Probability and Statistics, Chemical Rubber Co., Cleveland, Ohio (1968).

35. B. S. Reddy, J. H. Weisburger, and E. L. Wynder, *J. Nat. Cancer Inst.* **52**, 507 (1974).

36. N. D. Nigro, D. V. Singh, R. L. Campbell, and M. S. Pak, *J. Nat. Cancer Inst.* **54**, 439 (1975).

37. S. M. Grundy, E. H. Ahrens, Jr., and T. A. Miettinen, *J. Lipid Res.* **6**, 397 (1965).

38. J. H. Cummings, *Gut* **14**, 69 (1973).

39. D. P. Burkitt, *J. Am. Med. Assoc.* **231**, 517 (1975).

40. B. S. Drasar and D. Irving, *Br. J. Cancer* **27**, 167 (1973).

41. D. Kritchevsky and J. A. Story, *J. Nutr.* **104**, 458 (1974).

42. J. M. Ward, R. S. Yamamoto, and J. H. Weisburger, *J. Nat. Cancer Inst.* **51**, 713 (1973).
43. A. Navarette and H. J. Spjut, *Cancer* **20**, 1466 (1967).
44. J. C. Cleveland, S. F. Litvak, and J. W. Cole, *Cancer Res.* **27**, 708 (1967).
45. D. Wittig, G. P. Wilder, and D. Ziebarth, *Arch. Geschwulstforsch.* **37**, 105 (1971).
46. R. L. Campbell, D. V. Singh, and N. D. Nigro, *Cancer Res.* **35**, 1369 (1975).
47. B. S. Reddy, J. H. Weisburger, T. Narisawa, and E. L. Wynder, *Cancer Res.* **34**, 2368 (1974).
48. T. Narisawa, N. E. Magadia, J. H. Weisburger, and E. L. Wynder, *J. Nat. Cancer Inst.* **53**, 1093 (1974).
49. B. S. Reddy, A. Mastromarino, and E. L. Wynder, *Cancer Res.* **35**, 3403 (1975).
50. B. S. Reddy, J. H. Weisberger, and E. L. Wynder, *J. Nutr.* **105**, 878 (1975).
51. T. Asano, M. Pollard, and D. C. Madsen, *Proc. Soc. Exp. Biol. Med.* **150**, 780 (1975).
52. N. D. Nigro, R. L. Campbell, D. V. Singh, and Y. N. Lin, *J. Nat'l. Cancer Inst.* (in press).

Chapter 8

FECAL STEROIDS IN THE ETIOLOGY OF LARGE BOWEL CANCER

M. J. Hill

Bacterial Metabolism Research Laboratory
Rear of Colindale Hospital
London, England

I. INTRODUCTION

Boyland (1) has estimated that more than 90% of human cancers are due to environmental chemical factors and are therefore preventable. Similarly Higginson (2) analyzed the cancer registry data selecting the lowest incidence in any country for a given site as the minimum "inevitable" level of that cancer; incidences over and above this minimum level were then postulated as being due to environmental factors and therefore preventable. On this basis he calculated that more than 90% of cancers in any country were preventable. Although the concept of cancer as a preventable disease is greeted somewhat skeptically by many scientists, the demonstration of the links with recognized environmental factors in those cancers which are an occupational hazard (e.g., bladder cancer in the dye industry, osteomas in the match industry) and the subsequent prevention of further such cancers in those industries give grounds for optimism. Further, the demonstrable link between lung cancer and smoking gives grounds for hope that this major hazard may also be preventable.

In this chapter, I will review the grounds for suspecting that colorectal cancer may also be preventable and that diet, bacteria, and bile acids play major roles in its etiology. The key link is between large bowel cancer and diet; from this a number of hypotheses emerge which are currently being investigated and which are receiving various levels of experimental support. One such hypothesis involves bacteria and bile acids and this will be discussed in some detail together with the supporting data. Finally, the possible measures that could be taken to reduce the incidence of the disease if this postulate is correct are discussed.

II. THE EPIDEMIOLOGY OF COLORECTAL CANCER

The incidence of colorectal cancer is high in northwest Europe, North America, and Australia but low in Africa, Asia, and South America (Table I). Most of the high incidence populations are of North European stock and so this distribution could be due to racial factors; however, the incidence of the disease in Black Americans is similar to that in the white population, and the incidence is high in Argentina and in Uruguay. Further, migrants from Japan (where the incidence of large bowel cancer is low) to California (where the disease is relatively common) rapidly achieve an incidence of the disease little different from that of the resident Californian population (3); similar studies of migrants from various parts of the world to the United States have produced similar results (4,5), as have studies of migrations within the United States (4). This indicates that the factors most important in the etiology of the disease are environmental. Further, these factors

TABLE I. Incidence Rates of Large Bowel Cancer (Age-Standardized) in Various Countries[a]

Country	Incidence rate/100,000/annum, men aged 35–64	
	Colon	Rectum
Nigeria (Ibadan)	2.8	3.1
Uganda (Kyandondo)	0.0	3.5
India (Bombay)	6.6	8.0
Japan (Miyagi)	5.0	8.1
Singapore	4.6	8.5
Uruguay	26.9	18.7
Venezuela	7.3	4.4
Colombia (Cali)	5.7	3.9
England (Birmingham region)	17.4	20.7
Denmark	17.1	20.8
Sweden	16.0	12.8
Finland	8.3	7.5
Australia	23.5	13.7
New Zealand	30.6	20.8
USA (Connecticut)	31.2	20.6
USA (Hawaii, Caucasian)	27.6	12.4 β
Canada (Manitoba)	29.8	15.2

[a] Data from Doll (6).

rapidly affect the development (or genesis) of the disease so that the major determinant of a person's risk of large bowel cancer is his current life style rather than anything in his earlier life (4).

Although it is evident that the environment is of great importance in the etiology of the disease, there are numerous examples of populations living in the same area but with different incidences of colorectal cancer. Thus in South Africa, the white, Black, colored (mixed race), and Indian populations have greatly different incidences of the disease and have greatly different life styles although they suffer from the same air pollution, solar radiation, etc. Similarly, the Mormons and Seventh Day Adventist populations living in the United States, experience much lower incidences of the disease than the rest of the population. It seems, therefore, that the environmental factors involved are cultural and intimate rather than those shared by the whole population of an area. Such a factor is diet.

A dietary role in the causation of large bowel cancer was suggested by Stocks and Karn (7) in 1933, and has been supported by epidemiological data in many studies. Although there is general agreement on the role of diet, there is no agreement on the component of the diet principally responsible—since fat (8,9), animal protein (9,10), meat (11,12) refined sugar (13), lack of unrefined carbohydrate (14,15), and various vitamins (16) have been implicated. In studies of populations, there is a very strong correlation between the incidence of the disease and the amount of dietary fat and meat, a much weaker one with refined sugar, and no correlation at all with dietary fiber (Table II). In retrospective case-controlled studies the dietary factors implicated are fat, meat, and legumes (note that this positive correlation with legumes contrasts with the postulated negative correlation with unrefined vegetables). There have been no prospective studies of diet and large bowel cancer reported to date.

Thus, although there is still considerable argument, there is a large body of evidence accumulating which implicates dietary meat, fat, or animal protein in the causation of large bowel cancer; these three dietary items are, of course, highly interrelated.

Studies of the diet have not revealed any component carcinogen or cocarcinogen which correlates with the incidence of large bowel cancer and which could therefore be implicated in the causation of the disease. Because of this, it has been postulated (17) that the carcinogen or cocarcinogen is produced *in situ* in the colon by bacterial action on some benign substrate. The postulate then goes on to suggest that the diet determines the amount of substrate and also determines the composition of the gut bacterial flora. A variant of this postulate suggests that the crucial factor determining the amount of carcinogen produced (and therefore the risk of colorectal cancer) is the rate of transit of the colonic contents and that this is related to the amount of dietary fiber; in the original proposal the crucial factor was

TABLE II. The Correlation Between Dietary Components
and the Incidence of Colon Cancer[a]

Dietary component	Correlation coefficient
Fat	
total	0.81
animal	0.84
bound	0.88
Protein	
total	0.70
animal	0.87
Refined sugar	0.32
Fiber	
total	0.02
potatoes and starchy foods	−0.07
nuts	0.07
fruit	0.22
cereal	−0.32

[a] Data from Draser and Irving (9).

thought to be the concentration of substrate and the numbers of relevant bacteria.

If the basic concept is correct, then it should be demonstrable that bacteria can produce carcinogens or cocarcinogens from bile acids. If the original hypothesis is correct then the organisms responsible for production of the carcinogen should be more numerous in high-risk populations, and feces of these people should also contain higher concentrations of bile acids. If the later variation of the hypothesis is correct then the degree of degradation of bile acids should be dependent on the rate of transit of the gut contents, and the transit time in high-risk populations should be much slower than that in low-risk populations. Finally, the hypothesis should be capable of explaining the epidemiology of the disease and also of generating new studies which can also be used to test the validity of the concepts. First, though, we need to be sure that bile acids can act as carcinogens or cocarcinogens.

III. BILE ACIDS AS CARCINOGENS OR COCARCINOGENS

Bile acids have long been suspected of playing a role in human carcinogenesis and these suspicions were strengthened in 1933 when Wieland

and Dane (18) showed that deoxycholic acid could be converted to the very potent carcinogen 20-methylcholanthrene. Interest lapsed when there was such little headway made in demonstrating mechanisms for the desaturation and aromatization of the bile acid nucleus *in vivo*. However in 1939 Von Ghiron (19) reported that deoxycholic acid itself was carcinogenic to the mouse skin and Badger *et al.* (20) obtained similar results. These latter experiments were later made suspect by the isolation of a carcinogen from the oily vehicle used, but the lack of tumors in the control animals treated with oil alone, indicate that deoxycholic acid was at least cocarcinogenic; similar results were obtained using similar methods by Salaman and Roe (21).

Following these experiments, Narisawa *et al.* (22) tested several bile acids for cocarcinogenicity in the mouse rectum. They gave the animals a small dose of dimethylhydrazine (enough to cause tumors in a small proportion of animals) then repeatedly introduced bile acid solutions by rectal intubation; in this way they demonstrated that deoxycholic and lithocholic acid were both cocarcinogenic to the rectal mucosa.

Using a different model, results have been obtained by Nigro *et al.* adding further support to the suggestion that bile acids are cocarcinogenic in the rat large bowel. They treated their animals with dimethylhydrazine or with azoxymethane, then fed them a diet supplemented with cholestyramine; the test animals had many more tumors (23) than the controls (fed no cholestyramine) and further, the extra tumors were all in the large bowel while the numbers in the small intestine were not increased. There are a variety of possible explanations for these results, one of which is that the bile acids bound to the cholestyramine were nevertheless still available to act as cocarcinogens. In further studies (24), the experiments were repeated except that, instead of feeding cholestyramine to the animals, the bile was surgically diverted into the colon (bypassing the small intestine) with the same result as before.

In addition to these animal studies, deoxycholic acid has been shown to be mutagenic in drosophila (25) and in bacteria (26). Thus there is a body of evidence indicating that bile acids are mutagenic in bacteria and insects and cocarcinogenic in the mouse skin and in the rat and mouse rectum. There is also some evidence that they may be involved in human colon cancer, since in studies of populations with varying incidences of large bowel cancer there is a correlation between the fecal bile acid concentration and incidence of the disease (27–29). Obviously there is a need for more studies on the cocarcinogenicity or even carcinogenicity of the bile acids; nevertheless there is enough evidence to indicate that the hypothesis relating bacteria, bile acids, and colon cancer should not be ruled out on these grounds.

IV. BACTERIAL METABOLISM OF BILE ACIDS IN RELATION TO COLON CANCER

The major degradative reactions by bacteria on bile acids are: (a) deconjugation of bile salts to release the bile acids by the action of cholanoylglycine hydrolase; (b) oxidoreduction of the hydroxyl groups at positions 3,7, and 12; (c) dehydroxylation at C-7. In addition to these major reactions, which certainly occur in the normal human large intestine, there are a number of minor reactions which have been demonstrated *in vitro* using human gut bacteria but which still have to be demonstrated *in vivo*. These include: (d) dehydrogenation of the C^{4-5} bond in conjugation with a 3-oxo-group by the action of 3-oxo-5β-cholanoyl Δ^4-dehydrogenase; (e) aromatization of ring A of 3-oxo-4-cholenic acid; (f) aromatization of rings A and B of 3-oxo-4,6-choladienoic acid.

A. Deconjugation of Bile Salts

The hydrolysis of conjugated bile acids to yield the free bile acids (Fig. 1) is carried out by a high proportion of strains of anaerobic genera (*Bacteroides* spp., *Bifidobacterium* spp., *Eubacterium* spp., *Clostridium* spp., and *Veillonella* spp.) and of the enterococci (*Streptococcus faecalis, Strep. faecium*, etc.), but not by enterobacteria (*Escherichia coli, Klebsiella, Proteus* spp., etc.), *Lactobacillus* spp., or the oral streptococci (*Strep. viridans* or *Strep. salivarius*) found in feces (Table III).

TABLE III. The Proportion of Strains of Various Genera Able to Deconjugate Bile Salts, Dehydrogenate the 3α-, 7α- and 12α-Hydroxyl Groups and to Dehydroxylate at the 7-Position of Cholic Acid

Organisms	Number per gram feces	Hydrolase	Dehydrogenase 3α	Dehydrogenase 7α	Dehydrogenase 12α	Dehydroxylase
E. coli	10^8	0	39	78	10	1
Strep. faecalis	10^6	93	29	81	10	11
Strep. salivarius	10^7	0	0	0	0	0
Strep. viridans	10^7	0	0	0	0	0
Lactobacillus spp.	10^7	0	0	0	0	0
Bacteroides fragilis	10^{11}	82	30	79	10	44
Bifidobacterium spp.	10^{11}	74	21	56	10	40
Clostridium spp.	10^6	94	47	87	10	34
Veillonella spp.	10^4	50	10	50	0	4

The header spanning "Percentage with bile degradative ability" covers the Hydrolase, Dehydrogenase (3α, 7α, 12α), and Dehydroxylase columns.

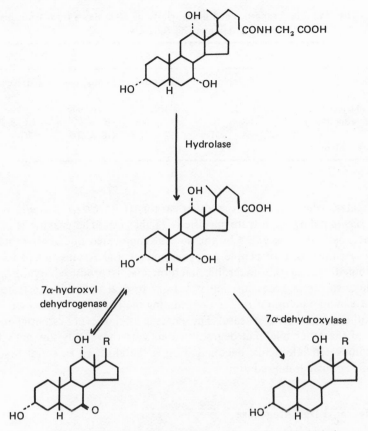

Fig. 1. The hydrolysis of glycocholic acid to release cholic acid which then can undergo 7α-hydroxydehydrogenation or 7α-dehydroxylation.

Cholanolyglycine hydrolase is a constitutive enzyme in all of the organisms studied and is cell-bound in *Bacteroides fragilis, Strep. faecalis,* and in *Clostridium welchii* while in *Bifidobacterium* spp. it is extracellular (30,31). The enzyme was first isolated in a purified state from the cytoplasm of *Cl. welchii* (32), and later from *B. fragilis, Bifidobacterium* spp and *Strep. faecalis* (31). The properties of the enzyme from various organisms are listed in Table IV.

The isolated purified enzyme has a *p*H optimum of 5–6 although in intact bacterial cells the optimum is 7. It is equally active under aerobic and under anaerobic conditions, and is not inhibited by substrate excess. Hydrolase from some organisms (e.g., *B. fragilis* and *Cl. welchii*) is equally active on conjugates of all bile acids and on both taurine and glycine

TABLE IV. The Properties of Cholanoylglycine Hydrolases from 4 Strains of
Human Gut Bacteria

	Strep. faecalis	Cl. welchii	B. fragilis	Bif. adolescentis
Location	Cytoplasm	Cytoplasm	Cytoplasm	Extracellular
pH optimum	4–5	5–6	5–6	5–6
Molecular weight		50–100,000	50–100,000	
Substrate specificity	GC TC	GC = TC	GC = TC	GC ≫ TC
	GC = GD	GC = GD	GC = GD	GC > GD
Product inhibition	–	+	+	–

conjugates, whereas that from other strains (e.g., *Strep. faecalis* or *Bif. adolescentis*) shows substrate specificity (Table IV). The enzyme is of high activity, as is demonstrated by the lack of conjugated bile acids in normal feces of either western people or of Ugandans and Asians; it is so widely distributed among the anaerobic bacteria and is produced in the gut in amounts so far in excess of that necessary to deal with the substrate load that it can play no major role in determining the risk of colorectal cancer or any other large bowel disease. The enzyme is, however, essential in that most of the other bile acid degradative enzymes have only low activity on the conjugated bile acids; deconjugation is therefore an essential first step permitting further degradation.

B. Hydroxyl Dehydrogenase

Hydroxycholanoyl dehydrogenases active an α-hydroxyl substituents at the 3, 7, and 12 positions have been characterized (33) and isolated (33–35). The enzyme active on the 7α-hydroxyl (Fig. 1) group is widely distributed among the genera of gut bacteria (Table III), that active on the 3α-hydroxyl somewhat less so, while that active on the 12α-hydroxyl group is present only in about 10% of strains; interestingly, of the bile acid degrading enzymes, the latter is the only one produced by a high proportion of strains (*E. Coli*) common to microbial biochemistry departments. The dehydrogenases have a requirement for NAD^+ or $NADP^+$, are cell-bound, and are readily isolated from the bacterial cytoplasm; they may have a pH optimum as high as 10 (for that acting on the 3α-hydroxyl group) or as low as 8 (for that acting on the 12α-hydroxyl group). They are reversible enzymes, producing the corresponding hydroxyl group optimally at neutral pH values, and are therefore oxidoreductases. The properties of the enzymes are summarized in Table V.

As with the cholanoyl hydrolase, the hydroxyoxidoreductases are widely distributed in bacteria isolated from feces of people living in areas with high or low incidences of colorectal cancer, and are therefore unlikely to be major determinants of the risk of this disease. However, the oxidation of the 3α-hydroxyl to a keto group is an essential reaction preparing the bile acid molecule for dehydrogenation of the steroid nucleus as described below.

C. Hydroxycholanoyl Dehydroxylase

Only the enzyme removing the 7α-hydroxyl group (Fig. 1) has been demonstrated (30,36) and isolated as a cell-free enzyme (33). The enzyme is almost entirely inducible, but is only produced when the culture medium has a pH higher than 6. It is produced by a high proportion of strains of anaerobic bacteria (the genera *Bacteroides, Bifidobacterium, Eubacterium, Clostridium,* and *Veillonella*) and of *Strep. faecalis* isolated from stools of western people (Table III). The enzyme produced by the anaerobic bacteria is only produced under extremely anaerobic conditions (much more exacting than those needed merely for luxurious bacterial growth) and has a pH optimum of 7–7.5. The enzyme has not been demonstrated to be reversible and, although it competes with the 7α-hydroxy oxidoreductase for the 7α-hydroxyl groups (the dehydrogenase being favored under aerobic conditions while the dehydroxylase is favored by anaerobic conditions) the outcome of this competition is usually the eventual removal of the substituent at a rate dependent on the redox potential. That the total enzyme activity in the gut

TABLE V. The Properties of Oxidoreductases Isolated from Various Strains of Gut Bacteria and Active on the 3α-, 7α-, or 12α-Hydroxyl Groups

	Enzyme active on the OH group at		
	3α	7α	12α
Inhibition by excess of substrate	ND	0–16%	ND
Enzyme activity (units/10^9 cells)	20–30	5–60	25
K$_m$ (mM)	0.1–0.3	0.1–0.5	0.2
pH optimum of oxidation reaction	10–11	9–10	8–8.5
of reduction reaction	6.5–8	6.5–7.5	6
Substrate specificity $\frac{\text{di subst}}{\text{tri subst}}$	1.2	1.2	1.1
Inducibility of the enzyme	++	++	++
Molecular weight	ND	50–100,000	ND

is high, is illustrated by the fact that although all bile acids synthesized by the liver have a 7α-hydroxyl group, only a small percentage of the normal fecal bile acids retain this substituent (37).

Samuelsson (38) showed in 1960 that the enzyme is in fact a dehydrase, removing both the 7α-hydroxyl and the 6β-hydrogen group; this trans-elimination reaction yields a Δ^6 intermediate (Fig. 2) which is unstable under intestinal conditions and is hydrogenated to the saturated product by nonenzymatic process. The formation of this unsaturated intermediate has been inferred from the results of studies using a doubly-labeled substrate but has never been isolated, or even detected, under *in vitro* or *in vivo* conditions; however, the enzyme does not have a requirement for saturated substrates with the 5β-configuration and consequently when the substrate with 4-en-3one configuration was used the Δ^6 bond was stabilized by conjugation (39,40) to give a product with a 4,6-dien-3-one structure.

Although the enzyme is produced by a high proportion of strains of anaerobic gut bacteria and of *Strep. faecalis* isolated from the stools of western people, such strains are relatively rare in the stools of populations with a low incidence of colon cancer (Table IV); the number of strains possessing the enzyme per gram of feces is related to the incidence of colorectal cancer (Fig. 3) as is the fecal concentration of the product of this

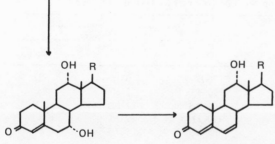

3-oxo-4,6-choladeinoic acid

Fig. 2. The 7α-dehydroxylation of cholic acid to yield deoxycholic acid via a Δ^6 intermediate; this can be stabilized in conjugation with a 4-en-3-one group to yield 3-oxo-4,6-choladienoic acid. R is the bile acid side chain C_4H_8COOH.

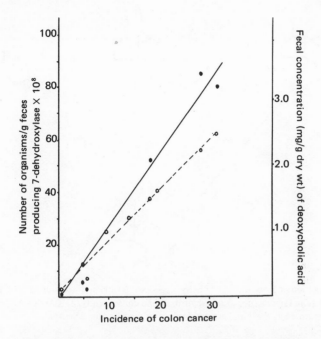

Fig. 3. The relationship between (a) the fecal concentration of deoxycholic acid and the incidence of colon cancer, and (b) the numbers of bacteria per gram feces producing 7α-dehydroxylase and the incidence of the disease. O----O, The fecal concentration of deoxycholic acid; ●----●, organisms producing 7α-dehydroxylase.

reaction, deoxycholic acid. The evidence indicating that this compound is carcinogenic or cocarcinogenic has already been discussed.

D. Nuclear Dehydrogenation Conjugated to a Keto Group

The three types of reaction discussed so far involved the substituents on the bile acid nucleus; the next three involve the nucleus itself and have in common the fact that they give rise to double bond formation. It should be noted, of course, that the dehydroxylation reaction also gives an unsaturated product which is stable under certain circumstances. The reaction of this type most readily carried out is the introduction of a double bond conjugated to a keto group (Fig. 4); two such reactions have been demonstrated using human gut bacteria and bile acids, the 3-oxo-cholanoyl Δ⁴-dehydrogenase and the 3-oxo-cholanoyl Δ¹-dehydrogenase. Since both

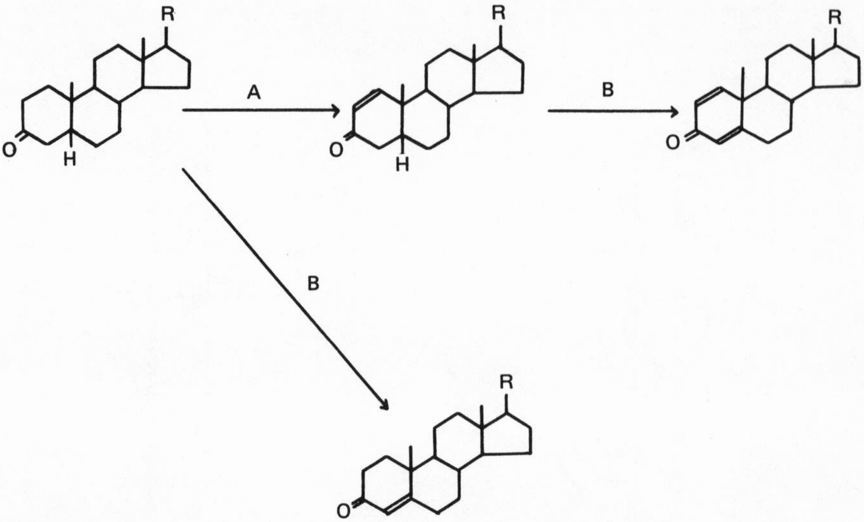

Fig. 4. The action of 3-oxo-steroid Δ¹-dehydrogenase (A) and of 3-oxo-steroid Δ⁴-dehydrogenase (B). Both require the presence of a 5β-hydrogen and consequently the 3-ox-1,4-choladienoic acid can only be formed by the initial action of the Δ¹-followed by the Δ⁴-enzyme.

enzymes have a requirement for a substrate with the 5β-configuration, introduction of a $C^{1\text{-}2}$ double bond can be followed by another at $C^{4\text{-}5}$ to give 3-oxo-chola-1,4-dienoic acid but prior action by the Δ⁴-enzyme prevents subsequent action by the Δ¹-dehydrogenase (41).

Both enzymes are inducible and are only produced by certain strains of clostridia (42) (Table VII). They require stringent anaerobic conditions for their production but can then act under aerobic conditions (using oxygen as the hydrogen acceptor) or under anaerobic conditions in the presence of a suitable hydrogen acceptor such as the nitrate reductase enzyme system, phenazine methosulfate, or quinones such as menadione of vitamin K (41,43). In the case of vitamin K, the whole molecule is required optimally, naphthaquinone or ubiquinone (Fig. 5) being much less active (43). The requirement for vitamin K need not be a limiting factor since we have demonstrated the presence of vitamin K in normal human feces (44) but its concentration appears to vary considerably from person to person. The properties of the Δ⁴-dehydrogenase are summarized in Table VIII.

Lecithinase-negative clostridia producing the Δ⁴-dehydrogenase are present in stools of about 40% of normal English people (17) and in those people, usually account for a high proportion of the total lecithinase-negative clostridia; these bacteria are relatively rare in the stools of Ugandans and Japanese (Fig. 6). To date we have no data on the amounts of the

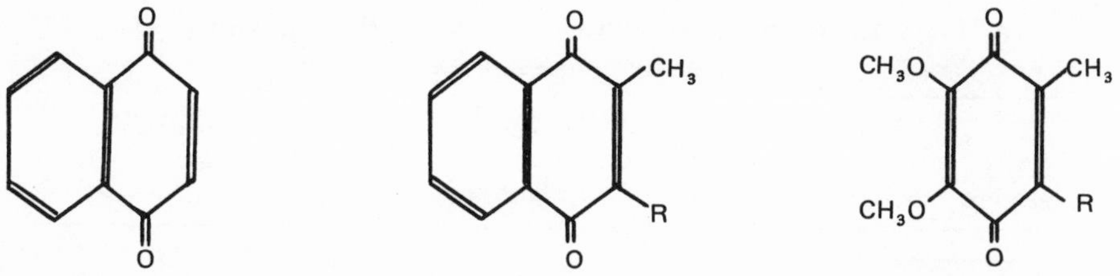

Naphthaquinone   Vitamin K   Ubiquinone

Fig. 5. The structure of vitamin K and its relationship to naphthaquinone and to ubiquinone.
$R = \left[CH_2-CH=C-CH_2 \atop \qquad\quad CH_3 \right]_n$ H. For vitamin K_2, n = 5–9; for ubiquinone, n = 6.

product of this reaction in stools of high- and low-incidence populations, although 3-oxo-4-cholenic acid has been reported in human feces (14).

Strains producing the Δ^4-dehydrogenase were isolated from stools of 82% of patients with colorectal cancer compared with only 43% of comparison patients in a recent case-comparison study (45), indicating that this enzyme may be involved in the etiology of colorectal cancer.

E. Aromatization of the A-ring of 3-oxo-4-Cholenic Acid

The essential step in the aromatization of the A-ring of the bile acid molecule is the removal of the methyl group at C-10. A proportion of the strains able to carry out the Δ^4-dehydrogenation were able to proceed to the aromatization of ring-A (Table VII) to give a phenolic steroid (42,46). When carried out by soil organisms (which are strictly aerobic) the removal of the C-10 methyl group is an oxidative reaction followed by the loss of

TABLE VI. The Proportion of 7α-Dehydroxylase-Producing
Strains from Feces of Various Populations

Population	Percentage of strains of anaerobic bacteria which produced dehydroxylase
England	42
Scotland	56
U.S.A.	50
Uganda	4
India	5
Japan	5

TABLE VII. The Proportion of Strains of Various Bacterial Groups Able to Produce Δ^4-Dehydrogenase and Aromatase

Organisms	Number tested	4-en-3-one from 3-oxo-5β-steroids (Δ^4-dehydrogenase)	Aromatic ring A from 4-en-3-one (aromatase)
Cl. paraputrificum	112	93	87
indolis	49	31	31
tertium	20	90	90
Cl. welchii	100	3	3
bifermentans	100	0	0
Others	314	8	4
E. coli	100	0	0
Strep. faecalis	100	0	0
Bacteroides spp.	100	0^a	0^a
Bifidobacterium spp.	75	0	0
Eubacterium spp.	25	0	0

^{a} Percentage of strains tested which were able to carry out the reaction.

CO_2 or formaldehyde. Clostrida are anaerobic organisms and it is not suprising, therefore, that they remove this group by a different mechanism; the reaction carried out by the clostridia involves removal of the 10-methyl group with an associated methylation at the periphery of the molecule (the 17-oxygen function in 4-androsten-3,17-dione or the carboxyl group of the bile acid analogue) as illustrated in Fig. 7. The removal of the methyl group from C-10 is accompanied by a C^{1-2} dehydrogenation and a subsequent keto-enolization to give the phenolic ring A. Unlike the enzyme from soil organisms, the 1,4-dien-3-one analogue could not act as a substrate, but the

TABLE VIII. Properties of the 3-Oxo-steroid Δ^4-Dehydrogenase from *Cl. paraputrificum*

Test	Property
Inducibility	+
Location of the enzyme	Cytoplasm
pH optimum	7–8
Substrate structural requirements	5β Configuration, presence of 3-oxo-group, absence of 7-oxo-group
Hydrogen acceptors in the absence of air	Menaphthone, phenazine methosulfate, vitamin K; but not NAD, NADP, FMN, FAD, or cytochrome c
Inhibitors of the reaction	Cu^{++}, Ca^{++}, formaldehyde, merthiolate, iodoacetate

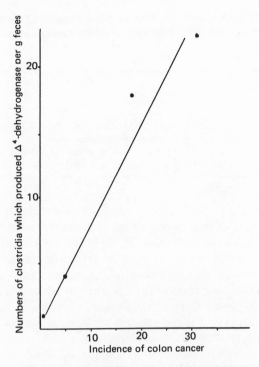

Fig. 6. The relationship between the numbers of clostridia which produced Δ⁴-dehydrogenase per gram of feces and the incidence of colon cancer.

enzyme from a small proportion of strains could utilize the 4,6-dien-3-one analogue. The properties of the enzyme are summarized in Table IX.

All of the strains able to aromatize ring A also produced the Δ⁴-dehydrogenase, but the converse was not true. No figures are available, but it is certain that strains able to carry out this reaction are much more common in stools of western people than in people from low-incidence countries. Similarly, it is extremely likely that the enzyme was present in the stools of

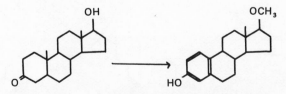

Fig. 7. The aromatization of 17β-hydroxy-4-androsten-3-one to yield 17β-methoxyestra-1,3,5(10)-trien-3-ol (46).

TABLE IX. Properties of the Enzyme(s) Produced by *Cl. paraputrifucum*
Converting 4-en-3-one Steroids to Their Phenolic Analogues (the Aromatase)

	Property
Inducibility	+
Location	Cytoplasm
pH optimum	8.5
Inhibitors	Cu^{++}, Ca^{++}, Mg^{++}, F^-, N_3^-, 10_4^-, iodoacetate, merthiolate, formaldehyde
Cofactors	NAD or NADP, phenazine methosulfate
Substrate requirements	4-en-3-one structure (not 1,4-dien-3-one)

a higher proportion of the colorectal cancer patients than of the comparison patients in the case-comparison study to which we already referred (45). Thus the evidence indicating a role for Δ^4-dehydrogenase in the etiology of colorectal cancer may also indicate a role of "aromatase." However, there is no data available concerning the presence or absence of phenolic bile acids in normal human feces although there is evidence that gut bacteria aromatize steroids *in vivo* in the intestine of the guinea pig (47).

F. Aromatization of Steroid Ring B

When 3-oxo-4,6-choladienoic acid was used as the substrate for the aromatization reaction described in the previous section the product was usually an analogue of equilin, having an aromatic ring A and one double bond in ring B; on occasions this product had undergone a further C^{8-9} dehydrogenation (39) to give an equilenin analogue with fully aromatic rings A and B (Fig. 8). This reaction is not readily reproducible even using 4,6-androstadien-3,17-dione as substrate (the androstane substrates are normally much easier to work with than the bile acids due to their greater solubility), indicating that we have still to find the optimal reaction conditions.

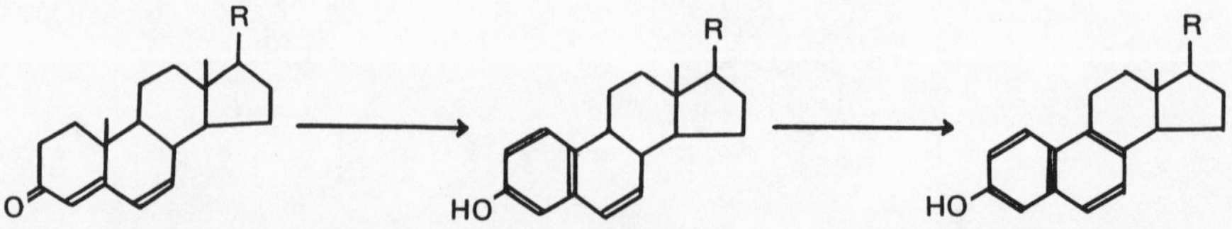

Fig. 8. The aromatization of ring A of a 4,6-androstadien-3-one to yield the estratetrene, then the estrapentene analogues (39).

At present this reaction is only of academic interest and there is no evidence concerning its relevance to the etiology of colorectal cancer.

G. Other Microbial Modifications of the Bile Acid Molecule

There are a range of other reactions which have been little studied. These include the desulfation of bile acid sulfate esters (48), and this reaction yields an unsaturated steroid product (Fig. 9). This reaction has been demonstrated *in vivo* and *in vitro,* but the organism(s) responsible has yet to be identified.

Further reactions include the formation of bile acid methyl and ethyl esters, and the modifications, via the enzymes already described, of the minor bile acids synthesized in the liver. These have not been studied extensively and their relevence to human colon cancer has not been investigated.

H. Summary

Bacteria modify the bile salt molecule at a number of points by a variety of reactions ranging from amide hydrolysis, desulfation, dehydroxylation, hydroxyl dehydrogenation, and nuclear dehydrogenation. Of these, the amide hydrolase and the hydroxyl dehydrogenases are so ubiquitous that they can play no role in a disease as relatively rare as colon cancer; the numbers of organisms per gram of feces producing 7α-dehydrolyase and Δ^4-dehydrogenase are highly correlated with the incidence of the disease while the relevence of other reactions has not been investigated.

V. THE RELATION OF MICROBIAL METABOLISM OF BILE ACIDS TO LARGE BOWEL CANCER

The postulate which has generated these studies (17,27,28) was that bacteria are able to metabolize bile acids to produce carcinogens or

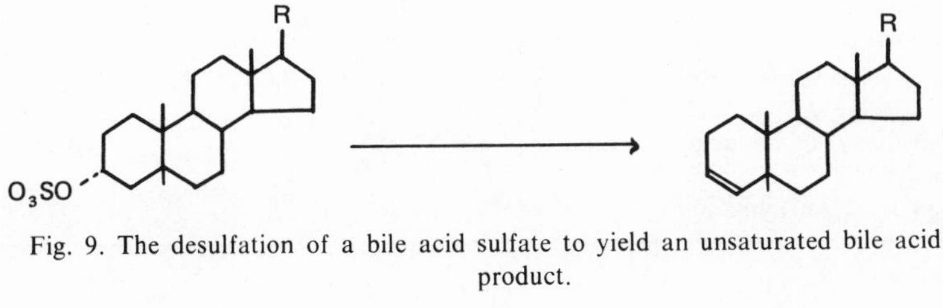

Fig. 9. The desulfation of a bile acid sulfate to yield an unsaturated bile acid product.

cocarcinogens and that these are important in the etiology of colon cancer. If this postulate is correct then we would expect that the incidence of colorectal cancer would be correlated with the activity of the enzymes involved in the production of the carcinogen and also with the concentration of the substrate for these reactions—the bile acids.

The relationship between fecal bile acid concentration and large bowel cancer has been supported by data from a range of sources. In a direct study of nine populations with varying incidences of the disease there was a good correlation between the incidence of the disease and the fecal concentration of bile acids, in particular of deoxycholic acid (27,28). These results have been further supported by indirect studies. Wynder and Reddy (29) studied the total fecal loss of bile acids in a number of populations living in New York City, and from their results it is clear that in new Japanese and Chinese migrants (in whom the incidence of the disease should be low) the fecal bile acid concentration was much lower than that in native-born Americans. Indians have a low incidence of the disease and their fecal level of bile acids is low (49). The incidence of large bowel cancer in black South Africans is much lower than that in the white population and the fecal bile acid concentration is also much lower (50). In Hong Kong the incidence of colorectal cancer increases with socio-economic class, and the fecal bile acid concentration of a group of high-income residents of high socio-economic class in Hong Kong was much greater (Table X) than that in a low-income group (51). Thus, there is now a large body of data indicating a relationship between the incidence of colorectal cancer and the concentration of bile acids in the gut contents.

If the postulate is correct, we would expect to find a greater activity against bile acids among bacteria from people living in high-risk areas than

TABLE X. Fecal Steroids in Three Income Groups Living in Hong Kong

	Group 1 (high income)	Group 2 (middle income)	Group 3 (low income)
Total neutral steroid concentration	5.90	5.12	4.13
Coprostanol + coprostanone cholesterol	0.60	0.54	0.55
Acids steroids			
Total acid steroid concentration (mg/g dry wt)	4.74	3.13	2.17
Total dehydroxycholanic acid concentration (mg/g dry wt)	1.65	1.16	0.90
% Mono+unsaturated bile acids	32	26	21

from those living in areas with a low risk of the disease. This has been demonstrated in two studies. Hill and Aries (37) showed that the fecal bile acids of British and American people were much more highly dehydroxylated than were those from Ugandan or Indian people; this was measured in terms of the proportion of mono- and unsubstituted bile acids in the total population (49% for English compared with 21% for Indian); these results were supported by the results of studies of three income groups in Hong Kong (Table X) where the high-income (and high-risk) group had a higher proportion of monosubstituted bile acids compared with the low-income group (who also had a lower risk of large bowel cancer). In support of this data, Wynder and Reddy (29) found a higher degree of degradation of steroids in their American subjects than in their Japanese or Chinese immigrants.

Since deoxycholic acid has been found to be a cocarcinogen in the mouse rectum, it was of interest to find that there was a good correlation between the fecal concentration of this bile acid and the incidence of colorectal cancer in nine populations (27). It has been estimated that in his first 50 years of life, the average Englishman passes 1.3 kg of deoxycholic acid in his stools (and therefore through his large intestine) and has a mean incidence of colon cancer of 18 per 10,000 per annum, so that deoxycholic acid would not need to be highly active to explain this risk.

Two bacterial enzymes are correlated with the risk of colorectal cancer, 7α-dehydroxylase and Δ^4-dehydrogenase. The 7α-dehydroxylase is the enzyme responsible for the formation of deoxycholic acid and therefore would be expected to be equally well correlated with the disease, but the Δ^4-dehydrogenase is totally unrelated, and there is further evidence which indicates that the correlation is not incidental.

If the postulate is correct, then those individuals who excrete high concentrations of bile acid and who carry the bacteria responsible for carcinogen production should be at much higher risk of colorectal cancer than those lacking one or other, or both, of these factors. This was tested in a retrospective study of 44 patients with colorectal cancer compared with patients admitted to the same wards with other gastrointestinal diseases (in this way the symptoms of the two groups were matched). Of the colorectal cancer patients 82% had a fecal bile acid concentration above a cut-off compared with 17% of the control patients (a fivefold difference); similarly 82% of the colorectal cancer patients had Δ^4-dehydrogenase-producing clostridia in their stools compared with 43% of the controls (a twofold difference). When these two criteria were combined 70% of the colorectal cancer patients had both high fecal bile acid levels and the relevant clostridia compared with only 9% of the controls (an eightfold difference).

Thus the possession of clostridia able to dehydrogenate the steroid nucleus is related to an increase of colorectal cancer and improves the discriminant value of the fecal bile acid concentration.

In the same study the discriminant value of the fecal deoxycholic acid concentration was examined. Of the colorectal cancer patients 14 of 14 had a fecal deoxycholic acid concentration above an arbitrary cut-off compared with 8 of 41 (or 19%) controls; 13 of the 14 colorectal cancer patients had the combination of high fecal deoxycholic acid and the nuclear dehydrogenating clostridia compared with 6 of the 41 (or 15%) controls. Thus the combination of high fecal deoxycholic acid–high clostridia characterizes a higher proportion of the cancer patients, but also a higher proportion of the controls, than the combination of high total bile acids–high clostridia.

Data from populations living in various parts of the world indicate a close correlation between the incidence of colorectal cancer and (a) the fecal concentration of bile acids, (b) the numbers per gram of feces of bacteria producing cholanoyl 7α-dehydroxylase, (c) the fecal concentration of deoxycholic acid (the product of the action of 7α-dehydroxylase), and (d) the numbers per gram of feces of bacteria producing 3-oxo-cholanoyl Δ^4-dehydrogenase. In studies of populations living within countries, but having different incidences of colorectal cancer, the population with the higher incidence also had a higher fecal concentration of total bile acids and of deoxycholic acid; unfortunately the bacterial enzymes were not assayed in these studies. Finally, in a retrospective study of colorectal cancer patients compared with patients with other gastrointestinal diseases, the risk of colorectal cancer was associated with high concentrations of fecal bile acids and of deoxycholic acid and with the presence of strains producing 3-oxo-cholanoly Δ^4-dehydrogenase.

All of these results add support to the thesis that bacterial degradation of bile acids produces a carcinogen or cocarcinogen in the large bowel. The next step must be to test the postulate in a prospective study.

In addition, it is necessary to identify the bile acids which are carcinogenic or cocarcinogenic. It has already been shown in a number of studies that deoxycholic acid has cocarcinogenic properties. The studies described above indicate that an unsaturated bile acid may be involved. The two enzymes correlated with the disease are the 7α-dehydroxylase and the 3-oxo-cholanoyl Δ^4-dehydrogenase; these two correlations, together with that of the total bile acids, suggest that the responsible bile acid might be 3-oxo-4,6-choladienoic acid (52) formed by the pathway described in Fig. 10. It is necessary, therefore, to identify the minor fecal bile acids and in particular to identify the unsaturated bile acids.

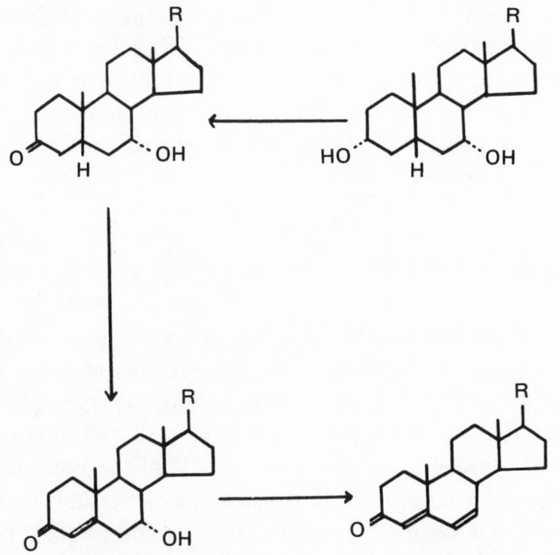

Fig. 10. The formation of 3-oxo-4,6-choladienoic acid from
chenodeoxycholic acid.

In addition, it is necessary to determine the factors controlling (a) the fecal bile acid concentration, (b) the numbers of bacteria producing 7α-dehydroxylase and 3-oxo-cholanoyl Δ^4-dehydrogenase, and (c) the activity and the production of those enzymes in the gut. Since the main factor in the etiology of the disease appears to be diet, its role in controlling these factors is of major importance. Although we need to know the answers to these questions in terms of the colon and rectum, in practice, only answers in terms of feces are practicable until adequate samples are available from, for example, colonic intubation.

VI. THE EFFECT OF DIET ON THE FECAL STEROIDS, BACTERIAL FLORA, AND GUT PHYSIOLOGY

Studies of the effect of diet on the fecal bile acid concentration are relatively easy to interpret. In general, these are reported in terms of total daily fecal loss of bile acids rather than the fecal concentration, but the qualitative effect of the dietary regime studied is often clear. In contrast, studies of the effect of diet on the intestinal bacterial flora are virtually impossible to interpret unless the enzymes of interest have been studied directly. Consequently there are data on the factors determining the fecal bile acid concentration and very few clues on the factors controlling the gut flora. There is an intermediate amount of data on the effect of diet on the gut luminal physiology.

A. The Effect of Diet on the Gut Bacterial Flora

The human gut bacterial flora is extremely complex, most known types of bacteria having been isolated from the intestine at some time or other. The flora can be divided into three groups (53): (a) those always present in large numbers (e.g., bacteroides, bifidobacteria), (b) organisms normally present in small numbers and apparently part of the normal resident flora (e.g., enterobacteria, enterococci), (c) organisms present in small numbers and not part of the normal resident flora but "contaminants" from elsewhere (e.g., staphylococci, bacilli, pseudomonads). Although the organisms in group (a) are numerically dominant they are mainly strictly anaerobic organisms, difficult to cultivate and little studied until the last 10–15 years. In contrast, the organisms in groups (b) and (c), although relatively unimportant numerically, are often easy to cultivate and have consequently been studied in considerable depth. The major intestinal organisms have already been listed in Table III.

Diet is usually considered to be of major importance in determining the composition of the gut bacterial flora. In the older literature it was reported that the chemical nature of the diet determined the predominant groups of bacteria present in the gut but although there is evidence for this in animals the results of studies in man have been equivocal (54,55). Differences are easier to demonstrate in the minor components of the flora; Crowther (56) found *Sarcina ventriculi* only in stools of vegetarians while Ueno *et al.* (57) found a Gram-negative anaerobic bacillus only in persons eating Japanese food. However, in studies of the effect of dietary meat (58), fat (59), bran (60), bagasse (sugar cane fiber, 60), and lactulose (61) there was no evidence that the numbers of organisms of the species responsible for 7α-dehydroxylase production had been altered. All of these studies were short-term—in most, the period of the test diet was four weeks—and this may not be long enough for the checks and balances in the gut flora to reach a new equilibrium. There is only one report of a long-term controlled dietary study (of a year or more); Hoffmann (62) found that when a high protein diet was eaten the flora was similar to that resulting from a balanced diet, while a high fat

diet favored the anaerobic bacteria (especially bacteroides) and a high carbohydrate diet favored the aerobic organisms. These studies were on a
single person and have not been repeated as yet. They must therefore be
treated with caution.

It is possible that such long-term studies may reveal significant differences only in a few individuals; although populations who have lived their
lifetime on different diets have differences in their gut bacterial flora, these
differences may be established in the early years of life. It is known that the
gut flora of young infants depends on whether they are fed breast or bottle
milk and it is possible that this factor, together with the diet in the initial
subsequent months, may determine the flora for life.

Nevertheless, although dietary manipulation has little immediate effect
on the composition of the gut bacterial flora there is evidence that it can
modify the enzymic activity of the flora. Volunteers had a higher fecal β-
glucuoronidase level when eating a high-meat diet than when eating a low-
meat diet (29) although the activity of this enzyme was not affected by the
amount of dietary bran or bagasse (K. Johnson, personal communication).
Similarly the fecal bile acids of people eating a high-fat diet are more extensively dehydroxylated than those of the same people on a low-fat diet (63).
There are conflicting reports on the effect on dietary fiber on intestinal 7α-
dehydroxylase; from the composition of the fecal bile acids, Walters *et al.*
(60) concluded that there was no effect, while Pomare *et al.* (64) concluded
from the composition of the biliary bile acids that bran reduced the activity
of the enzyme.

From these studies it is evident that only studies of the direct effect of
diet on the enzymes of interest rather than its effect on the composition of
the bacterial flora *per se* are likely to yield useful results.

B. The Effect of Diet on the Fecal Bile Acid Concentration

Many of the studies of the factors determining the fecal loss of bile
acids have been carried out on patients with abnormal lipid metabolism fed
liquid-formula diets. Until the effects of the solid indigestible components
of the normal diet are more fully understood the conclusions to be drawn
from the results of such studies will not be clear. At present we can only
interpret studies of normal people eating normal solid diets; very few studies
have been reported.

Probably the best such study is that by Antonis and Bersohn (50) on
white and Black South African prisoners. Two basal diets were used. One
approximated that of the urban Bantu population and contained 15% fat

calories, 15% protein calories, 70% carbohydrate calories, and 15 g fiber/
day; the second approximated that of the urban white population and
contained 40% fat calories, 15% protein calories, 45% carbohydrate
calories, and about 5 g fiber/day. There were intermediate diets (including
high fat–high fiber and low fat–low fiber) and there were variations in the
source of the fat (including butter and sunflower oil). The dietary periods
varied from 8–39 weeks. In those on a low fiber diet (Table XI), an increase
in the amount of dietary fat resulted in an increased fecal concentration of
bile acids (with the fat from both butter and sunflower oil) and of neutral
steroids (with sunflower oil but not with butter). In those on a high fiber
diet, an increase in the amount of fat resulted in a small increase in the fecal
bile acid concentration if the fat was from sunflower oil but not if butter fat
was used. Similar results on the fecal bile acid concentration (Table XI)
were obtained in a short term study (63) in which the amount of dietary fat
from all sources was decreased from the normal (100–120 g/day) to less
than 30 g/day. The amount of animal protein was kept constant; fat meat
such as beef and pork being replaced by lean meat (such as chicken) and
fish; the amount of roughage was not increased greatly (this was reflected in
the unchanged mean stool weight) and caloric intake was maintained by
increasing the intake of refined carbohydrate.

The main shortcoming of these studies is that they were relatively
short-term and give no information on long-term effects. There are many
dietary intervention studies being carried out by those interested in coronary
heart disease; these involve changes in dietary fat intake and are intended to

TABLE XI. The Effect of Reducing the Amount of Dietary Fat on the Fecal Steroid
Concentration

	Number of subjects	Duration of diet (weeks)	Fecal steroid concentration as % of the control value	
			Acid steroids	Neutral steroids
Low fiber				
High fat [G] → low fat [a]	4	1–4	28%	91%
High fat [O/B] → low fat [b]	25	18	64%	92%
High fiber				
High fat [B] → low fat [b]	9	22	109%	47%
High fat [O] → low fat [b]	17	22	83%	24%

[a] Reference (63).
[b] Reference (50).

TABLE XII. The Effect of Daily Supplements of Dietary Fiber on Fecal Steroid Analyses

Diet[b]	Number of subjects	Duration of diet (weeks)	Fecal steroid concentration as % of the control value[a]	
			Acid steroids	Neutral steroids
100 g added bran	4	3	57	56
39 g added bran	4	4	63	63
16 g added bran	8	3	61	?
10.5 added bagasse	10	12	100	61
Vivonex[c]	3	2	45	24

[a] The effect on the fecal steroid concentration of adding fiber to the diet. The results for the high fiber diet are expressed as a percentage of the control (no added fiber) values.
[b] Meat and fat held constant.
[c] Vivonex contains no fiber and very little fat. The protein is in the form of amino acids and small peptides.

last many years and so would be ideal in determining the long-term effect of dietary fat on fecal bile acids.

Replacement of dietary meat by vegetable protein normally results in a decreased fat intake. Therefore, the results of studies of English vegans (65), of New York vegetarians and Seventh Day Adventists (29), or the effect of dietary meat in uncontrolled studies (66), in all of which the people on a no-meat diet had a decreased fecal bile acid concentration, could readily be explained as the result of decreased fat intake.

There have been a number of studies relating the fecal bile acid concentration to the amount of dietary fiber. Again, all of these are fairly short-term; the longest was that of Antonis and Bersohn (50) where the diets lasted up to 39 weeks. They found that in persons consuming a high-fat diet the increase in the dietary fiber from 4 to 15 g resulted in a 58% decrease in fecal bile acid and 34% decrease in fecal neutral steroid concentration while in those consuming a low-fat diet the same change in fiber intake resulted in reductions in the fecal concentrations of 44% in the acid and 16% in the neutral steroids.

In three different short-term (3–4 weeks) studies the addition of 16,39, and 100 g bran to the diet (60,67,68) all produced a 40% decrease in fecal acid steroid concentration (Table XII) since the increase in fecal bulk was greater than the increased daily loss of bile acids. This was not so when the dietary fibre used was bagasse; this had a bulking action but also appeared to bind bile acids since there was an equal increase in their fecal loss resulting in an unchanged fecal bile acid concentration. The role of dietary fiber on bile acid metabolism is discussed in detail in Chapter 9 of this volume.

C. The Effect of Diet on the Activity of Bacterial Enzymes in the Gut

Diet might affect the activity of bacterial enzymes by (a) speeding the rate of transit and so reducing the time available for bacterial metabolism to take place, (b) changing the colonic pH, (c) changing the colonic E_h, (d) other mechanisms.

To date we have no evidence that the rate of transit of material through the gut is important in determining the degree of degradation of colonic substrates; Fig. 11 relates transit time to the degradation of cholesterol (lysine and arginine behave similarly); in these there is no relation between transit time and extent of metabolism. Indeed in volunteers consuming a soluble defined diet the neutral steroids were virtually undegraded (69) even though the transit time in people on such a diet is about 2 weeks (or about 5 times the normal length of time). A possible explanation for this has been suggested (70) and is illustrated in Fig. 12. When the time of transit is less than A there is virtually no time for degradation of substrates, when the time taken is between A and B hours the extent of degradation is

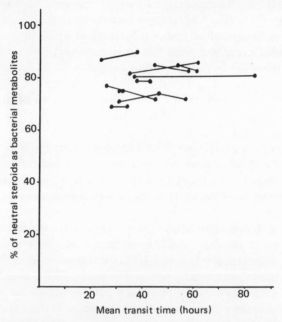

Fig. 11. The relation between the mean transit time of gut contents and the degree of degradation of cholesterol to the bacterial metabolites coprostanol and coprostanone.

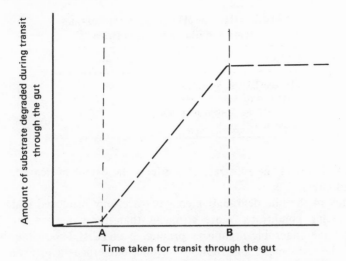

Fig. 12. The relation between transit time and metabolism. When the transit time is less than A hours, material passes through the gut too rapidly for metabolism to occur; between A and B hours, the amount of metabolite produced is dependent mainly on the time available; when the contents take more than B hours to pass through the gut maximal metabolism has taken place and incremental increases in time result in no further metabolism.

determined by the amount of time available while if the transit time is longer than B hours there is no extra degradation because the reaction has gone to completion. If the bacterial enzyme is of high activity, times A and B will be short—shorter than the transit time that can be achieved by the use of dietary fiber—and presumably this is the case with the enzymes responsible for secondary amine production and for cholesterol metabolism.

Most of the enzymes involved in bile acid metabolism have pH optima close to neutrality (Table XIII) and are inhibited by acid conditions *in vitro*, the only exception being cholanoyl hydrolase. Ugandans, whose mean fecal pH was 5–7 had fecal acid steroids that had undergone much less dehydroxylation that those in English feces (37) which had a mean pH of 6.7. It is possible to reduce the intestinal pH by the use of such compounds as lactulose (which gives rise to fermentative diarrhea) or magnesium sulfate (71). Both of these compounds give rise to a more rapid transit of the bowel contents and also have a bulking action in addition to their effect on the pH; it will be important to separate these three mechanisms when studies of the effect of these agents on bile acid degradation are carried out

TABLE XIII. The pH Optima of the Enzymes
Involved in Bile Acid Degradation

Enzyme	pH optimum
Cholanoylglycine hydrolase	5–6
7α-Dehydroxylase	7–7.5
3-Oxo-steroid Δ⁴-dehydrogenase	7–8
Aromatase	8.5

because dilution of the substrate also effects the extent of degradation (as described later).

Many of the bile degradative enzymes are only produced under stringent reducing conditions, more stringent than those needed merely for growth of the anaerobic organisms producing them. It is possible that the ratio of the number of anaerobic to aerobic bacteria is a measure of the anaerobiosis of the gut and this ratio is related to the degree of dehydroxylation of the fecal bile acids (Fig. 13). The proportion of aerobic bacteria in the total flora is highest when the diet contains high carbohydrate levels, lowest when a high fat diet is consumed, and intermediate when a high protein or a balanced diet is consumed (62).

The Δ⁴-dehydrogenase has a requirement for a suitable hydrogen

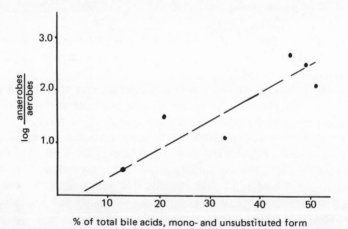

Fig. 13. The relation between the $\log_{10}$ number of anaerobic bacteria/ number of aerobic bacteria and the proportion of the total bile acids in the form of mono- and unsubstituted steroids. The former is a measure of intestinal E_h while the latter is a measure of the dehydroxylation undergone by the bile acids.

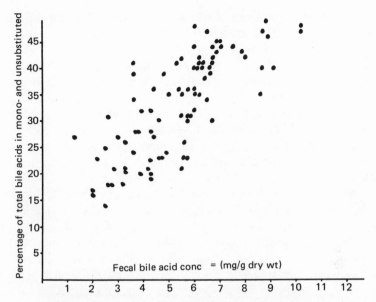

Fig. 14. The relation between the total fecal bile acid concentration and the proportion in the form or mono- and unsubstituted bile acids.

acceptor cofactor, the most likely being vitamin K which is a very potent cofactor for this enzyme and is present in the human gut (44). There are no data on the factors determining the concentration of vitamin K in the gut except that it was unaffected by the addition of 39 g bran for three weeks or of 16 g bagasse for twelve weeks (44).

Among the other factors affecting the metabolism of steroids by gut bacteria the most relevant is dilution. The extent of dehydroxylation of the fecal bile acids is related to the fecal bile acid concentration (Fig. 14). The enzyme 7α-dehydroxylase is inducible and perhaps in the gut the enzyme is only fully induced when the bile acid concentration is relatively high. At concentrations above 6 mg/g dry weight there appears to be no further increase in the extent of dehydroxylation; a downturn in the graph at very high concentrations would be expected, since the enzyme is inhibited *in vitro* by substrate excess (33).

VII. CONCLUSIONS

In this chapter the epidemiology of large bowel cancer is described; the consensus opinion is that diet is strongly implicated in the etiology of the disease. There is no consensus concerning the dietary component responsible, but the two that have received most attention are fiber depletion and fat excess; mechanisms have been suggested to explain both.

The data linking the intake of fiber to the incidence of the disease has been reviewed; it is insubstantial and there is much more evidence to indicate that the two are *not* linked. Further, the suggested mechanism by which fiber might protect against the disease has been examined and shown to be extremely unlikely. However, in the latter part of the chapter it is indicated that one effect of a high-fiber diet might be to dilute the fecal contents (in particular, the fecal cocarcinogens) and thereby reduce the risk of the disease.

A mechanism has been postulated to rationalize the link between dietary fat and large bowel cancer; this involves the production of a cocarcinogen by gut bacteria from the bile acids. The evidence in favor of this has been summarized; it consists of the results of studies of populations with varying risk of the disease and of case-control study of large bowel cancer patients compared with patients with other gastrointestinal diseases. Although these studies are at an early stage there are grounds for suspecting that they might be on the right lines and that the etiology of this major disease might be on the point of being discovered. If the postulated mechanism proves to be correct then it would mean that the disease is probably preventable by dietary manipulation. How this might be achieved has also been discussed; it is earlier to be optimistic about our ability to discover the etiology of large bowel cancer than about our ability to persuade the general population to make the changes in their life style necessary to prevent the disease.

REFERENCES

1. E. Boyland, *The Practioner* **199**, 277 (1967).
2. J. Higginson, *Proc. R. Soc. Med.* **61**, 723 (1968).
3. P. Buell and J. E. Dunn, *Cancer* **18**, 656 (1965).
4. W. Haenszel and E. A. Dawson, *Cancer* **18**, 265 (1965).
5. J. Staszewski and W. Haenszel, *J. Nat. Cancer Inst.* **35**, 291 (1965).
6. R. Doll, *Br. J. Cancer* **23**, 1 (1969).
7. P. Stocks and M. K. Karn, *Ann. Eugen.* (London) **5**, 237 (1933).
8. E. L. Wynder and T. Shigematsu, *Cancer* **20**, 1520 (1967).
9. B. S. Drasar and D. Irving, *Br. J. Cancer* **27**, 167 (1973).
10. O. Gregor, R. Toman, and F. Prusova, *Gut* **10**, 1031 (1969).
11. B. Armstrong and R. Doll, *Int. J. Cancer* (1975).
12. W. Haenszel, J. W. Berg, M. Segi, M. Kurihara, and F. B. Locke, *J. Nat. Cancer Inst.* **51**, 1765 (1973).

13. T. L. Cleave, *The Saccharine Diseases,* Wright (Bristol) (1974).
14. A. R. P. Walker, B. F. Walker, and B. D. Richardson, *Br. Med. J.* iii, 238 (1969).
15. D. P. Burkitt, *J. Nat. Cancer Inst.* 47, 913 (1971).
16. E. Bjelke, *Scand. J. Gastroenterol.* 9, suppl. 31 (1974).
17. V. C. Aries, J. S. Crowther, B. S. Drasar, M. J. Hill, and R. E. O. Williams, *Gut* 10, 334 (1969).
18. H. Wieland and E. Dane, *Z. Physiol. Chem.* 219, 240 (1933).
19. V. Ghiron, *Proc. 3rd. Int. Cancer Congr.,* p. 116 (1939).
20. G. M. Badger, J. W. Cook, C. L. Hewett, E. L. Kennaway, N. M. Kennaway, R. H. Martin, and A. M. Robinson, *Proc. R. Soc. London, Ser. B.* **129,** 439 (1940).
21. M. H. Salaman and F. J. C. Roe, *Br. J. Cancer* 10, 363 (1956).
22. T. Narisawa, H. Nakano, M. Hayakawa, T. Sato, and A. Sakuma, *in* "Topics in Chemical Carcinogenesis" (W. Nakahara *et al.,* eds.), p. 145, Baltimore, University Park Press (1972).
23. N. D. Nigro, N. Bhadrachari, and C. Chomchai, *Dis. Colon Rectum* 16, 438 (1973).
24. C. Chomchai, N. Bhadrachari, and N. D. Nigro, *Dis. Colon Rectum* 17, 310 (1974).
25. K. A. Jensen, I. Kirk, G. Kolmare, and M. Westergaard, *Coldspring Harbor. Symp. Quant. Biol.* **16,** 245 (1951).
26. M. Demerec, *Br. J. Cancer* 2, 114 (1948).
27. M. J. Hill, B. S. Drasar, V. C. Aries, J. S. Crowther, G. Hawksworth, and R. Williams, *Lancet* i, 95 (1971).
28. M. J. Hill, and B. S. Draser, *in* "Anaerobic Bacteria: Role in Disease" (A. Balows, R. M. De Haan, V. R. Dowell, and L. B. Guze, eds.), p. 119, Thomas (Springfield) (1974).
29. E. L. Wynder and B. S. Reddy, *Cancer* 34, 801 (1974).
30. M. J. Hill and B. S. Drasar, *Gut* 9, 22 (1968).
31. V. C. Aries and M. J. Hill, *Biochim. Biophys. Acta* **202,** 526 (1970).
32. P. P. Nair, M. Gordon, and J. Reback, *J. Biol. Chem.* **242,** 7 (1967).
33. V. C. Aries and M. J. Hill, *Biochim. Biophys. Acta* **202,** 535 (1970).
34. I. A. MacDonald, N. Williams, D. E. Mahoney, and W. M. Christie, *Biochim. Biophys. Acta* **384,** 12 (1975).
35. G. A. D. Haslewood, G. M. Murphy, and J. M. Richardson, *Clin. Sci.* **44,** 95 (1973).
36. T. Midtvedt, *Acta Path. Microbiol. Scand.,* **71,** 147 (1967).
37. M. J. Hill and V. C. Aries, *J. Pathol.* **104,** 129 (1971).
38. B. Samuelsson, *J. Biol. Chem.* 235, 361 (1960).
39. P. Goddard and M. J. Hill, *Trans. Biochem. Soc.* 1, 1113 (1973).
40. S. Hayakawa, Y. Kanematsu, and T. Fukiwara, *Biochem. J.* **115,** 249 (1969).
41. V. C. Aries, P. Goddard, and M. J. Hill, *Biochim. Biophys. Acta* **248,** 482 (1971).
42. P. Goddard, F. Fernandez, B. West, M. J. Hill, and P. Barnes, *J. Med. Microbiol.* **8,** 429 (1975).
43. F. Fernandez and M. J. Hill, *J. Med. Microbiol.* **8,** p. ix (1975).
44. F. Fernandez and M. J. Hill, In preparation.
45. M. J. Hill, B. S. Drasar, R. E. O. Williams, T. W. Meade, A. G. Cox, J. E. P. Simpson, and B. C. Morson, *Lancet* i, 535 (1975).
46. P. Goddard and M. J. Hill, *Biochim. Biophys. Acta* **280,** 336 (1972).
47. P. Goddard and M. J. Hill, *J. Steroid Biochem.* 5, 569 (1974).
48. A. Norman and R. H. Palmer, *J. Lab. Clin. Med.* 63, 986 (1964).
49. K. S. Shurpalekar, T. R. Doraiswamy, O. Sundaravalli, and M. Narayana Rao, *Nature* 232, 554 (1971).
50. A. Antonis and I. Bersohn, *Am. J. Clin. Nutr.* 11, 142 (1962).

51. M. J. Hill, *Cancer* **34**, 815 (1974).
52. M. J. Hill, *Cancer* **36**, 2387 (1975).
53. B. S. Drasar and M. J. Hill, *in* "Human Intestinal Flora," Academic Press, London (1974).
54. H. Haenel, W. Muller-Beuthow, and A. Schuenert, *Zentralbl. Bakteriol. Parasitenk. Infektionsk. Hyg. Abt. 1 Orig.* **168**, 37 (1957).
55. W. E. C. Moore, E. P. Cato, and L. V. Holdeman, *J. Infect. Dis.* **119**, 641 (1969).
56. J. S. Crowther, *J. Med. Microbiol.* **4**, 343 (1971).
57. K. Ueno, P. T. Sugihara, K. S. Bricknell, H. R. Atterbury, V. L. Sutter, and S. M. Finegold, *in* "Anaerobic Bacteria: Role in Disease" (A. Balows, R. M. DeHaan, V. R. Dowell, and L. B. Guze, eds.), p. 135 (1974).
58. B. R. Maier, M. A. Flynn, G. C. Burton, R. K. Tsutakawa, and D. J. Hentges, *Am. J. Clin. Nutr.* **27**, 1470 (1974).
59. J. S. Crowther, Ph.D. Thesis, University of London (1971).
60. R. Walters, I. M. Baird, P. S. Davies, M. J. Hill, B. S. Drasar, D. A. T. Southgate, J. Green, and B. Morgan, *Br. Med. J.* **2**, 537 (1975).
61. A. Vince, R. Zeegan, J. E. Drinkwater, F. O'Grady, and A. M. Dawson, *J. Med. Microbiol.* **7**, 163 (1974).
62. K. Hoffmann, *Zentralbl. Bakteriol. Parasitenk. Infectionsk. Hyg. Abt. 1 Orig.* **192**, 500 (1964).
63. M. J. Hill, *J. Pathol.* **104**, 239 (1971).
64. E. W. Pomare, K. W. Heaton, T. S. Low-Beer, and S. White, *Gut* **15**, 824 (1974).
65. V. C. Aries, J. S. Crowther, B. S. Drasar, M. J. Hill, and F. R. Ellis, *J. Pathol.* **103**, 54 (1971).
66. B. S. Reddy, J. H. Weisburger, and E. L. Wynder, *Science* **183**, 416 (1974).
67. M. A. Eastwood, J. R. Kirkpatrick, W. D. Mitchell, A. Bone, and T. Hamilton, *Br. Med. J.* **4**, 392 (1973).
68. D. J. A. Jenkins, J. H. Cummings, and M. J. Hill, *Am. J. Clin. Nutr.* **28**, 1408 (1975).
69. J. S. Crowther, B. S. Drasar, P. Goddard, M. J. Hill, and K. Johnson, *Gut* **14**, 790 (1973).
70. M. J. Hill, *Digestion* **11**, 289 (1974).
71. R. L. Bown, J. A. Gibson, G. E. Sladen, B. Hicks, and A. M. Dawson, *Gut* **15**, 999 (1974).

Chapter 9

DIETARY FIBER AND BILE ACID METABOLISM*

David Kritchevsky and Jon A. Story

The Wistar Institute of Anatomy and Biology
Philadelphia, Pennsylvania

In recent years, the role of nonnutritive fiber in the diet and its effects on various metabolic parameters has assumed considerable importance. Epidemiological studies have suggested that populations subsisting on high fiber diets are relatively free from certain diseases, such as colon cancer and coronary heart disease, that are common in the Western world (1,2). It is not the intent of this exposition to discuss differences in lifespan, general health, and cause of death between the economically over- and under-developed areas of the world. However, it is important to point out that the current interest in fiber derives to a large extent from the publicity attending the epidemiological observations cited above.

In the ensuing discussion, many different substances will be discussed. Although they all fall under the general classification of fiber, they are structurally different. Spiller and Amen (3) have pointed up the need for better identification of plant fibers. Fiber consists of cellulose and hemicellulose (both of which are carbohydrate in nature) and lignin which is the designation for a complex mixture of phenylpropane polymers. Eventually, all reports of work with dietary fiber will have to contain a detailed analysis of the particular substance under study.

Dietary fiber appears to exert an effect on lipid metabolism in animals and man. In general, studies of this effect have concentrated upon changes in serum lipids. There have been efforts to explain the mechanism(s) of the

* Supported, in part, by USPHS research grants HL-03299, HL-05209 and a Research Career Award HL-0734 from the National Heart and Lung Institute and by grants-in-aid from the National Dairy Council and the National Live Stock and Meat Board.

hypolipidemic effect of some types of fiber and most of these have concerned themselves with the influence of fiber on bile acid metabolism. The hypolipidemic effects of fiber have been reviewed recently (4) and this exposition will discuss them only insofar as they are mediated by way of changes in bile acid metabolism.

Portman and his coworkers (5,6) found that diet affected the bile acid spectrum and output in bile duct cannulated rats. In general, commercial laboratory ration resulted in a twofold increase in total bile acid excretion. To ascertain that the diet and not the cannulation was responsible for the observed effect, he repeated the feeding experiment (7) and determined cholic acid half-life and the excretion of digitonin-precipitable sterols and cholic acid. The results (Table I) indicate that on a semipurified diet, there is decreased steroid excretion. The addition of chow lipid or chow sterols to the semipurified, sucrose diet had little effect on cholic acid half-life or excretion; substitution of lactose for sucrose increased cholic acid half-life by 55% and reduced excretion by 27% (8).

Other experiments in rats have indicated that various types of fiber affect steroid dynamics and excretion. The exact patterns of excretion differ, however, which may reflect differences in mechanisms of action. Thus, Leveille and Sauberlich (9) fed rats 1% cholesterol with and without pectin (5%). The serum and liver cholesterol levels were significantly reduced in the latter group. The pectin-fed rats did not excrete any more sterol than did the cholesterol-fed controls (possibly because of the level of dietary cholesterol) but they excreted 32% more bile acids. Rats fed psyllium seed hydrocolloid (10) exhibited reduced bile acid turnover time.

Kritchevsky *et al.* (11) fed rats special diets which contained cellulose and differed in caloric distribution and compared them with rats fed a stock diet. Two days after the rats had been fed one dose of [4-^{14}C]cholesterol, the

TABLE I. Steroid Excretion and Cholic Acid Half-Life in Rats Fed Various Diets[a]

Diet	No.	Cholic acid $t_{1/2}$ (days)	Fecal steroids, mg/kg/day	
			Cholic	Digitonin-precipitable
Laboratory	4	2.00 + 0.23 a	36.4 + 4.8 xy	75.4
S[b]-starch	4	3.24 + 0.53 b	10.3 + 1.4 x	48.1
S-sucrose	3	4.17 + 0.53 ac	7.7 + 0.9 yz	44.0
S-sucrose-C[c]	4	1.44 + 0.21 bc	23.4 + 4.0 z	29.9

[a] After Portman and Murphy (7). Values bearing same letter are significantly different.
[b] 20% Casein, 8% corn oil, 67.6% starch or sucrose.
[c] 20% Celluflour added.

TABLE II. Effect of Fiber Source on Cholesterol Absorption in Rats[a]

| Diet[b] | Fiber | Wt. gain, g | Radioactivity | |
			Serum plus liver dpm $\times$ 10^4	Feces, dpm $\times$ 10^5
Dextrose	Cellulose	125	3.39	2.77 (27)[c]
	Alfalfa	99	2.02	4.29 (15)
Sucrose	Cellulose	125	2.58	2.94 (15)
	Alfalfa	107	2.10	5.00 (17)
Corn oil	Cellulose	115	2.83	2.63 (22)
	Alfalfa	86	1.21	4.52 (17)
Casein	Cellulose	134	4.11	2.11 (31)
	Alfalfa	98	3.31	3.86 (35)

[a] After Kritchevsky et al. (12).
[b] Diets contained 50% of calories as component listed with other 50% divided evenly between other two sources of calories.
[c] % As acidic steroid.

control rats excreted 300% more radioactive acidic steroid than did any of the test groups. In a second experiment (12), rats were fed similar isocaloric, isogravic diets in which the fiber was either cellulose or alfalfa. Weight gain was good in all groups but within each pair of diets, weight gain was higher in the rats fed cellulose. Serum and liver cholesterol levels did not differ greatly in the two groups but, the alfalfa-fed rats absorbed less and excreted more of a single dose of [4-^{14}C]cholesterol. The results are summarized in Table II. The relative amount of acidic steroid was considerably higher in both high protein groups. Chickens fed a diet containing 43% corn starch excrete 33% more lipid (25% more sterol) than when the diet contains 43% ground oats (13).

Antonis and Bersohn (14) compared different diets in White and Bantu prisoners in South Africa and found that when the diet contained more fiber, the stools were bulkier and contained more bile acids, fatty acids, and sterols (Table III). The Bantu diet, which normally contains a high level of fiber, is reported to decrease bowel transit time by as much as 50% when compared to the average White diet in South Africa (15,16).

Other studies in man have shown similar effects of fiber. Some studies compared vegetarian with mixed diets whereas others fed specific substances. In one study, two healthy young volunteers were fed 9.6 g of psyllium seed hydrophilic colloid daily for six weeks. There was no change in their fecal neutral steroid excretion, but fecal bile acid excretion was increased almost threefold (17).

In a comparison of subjects consuming mixed and strict vegetarian

TABLE III. Influence of Dietary Fiber Level on Fecal Lipids in Man[a]

Diet	Group[b]	Fiber, g	Bulk	Bile acids	Sterols	Fatty acids
				Daily excretion, g		
A	W	13.3	275	0.48	0.20	2.06
B	B	16.2	311	0.51	0.15	3.77
C	W	11.5	234	0.45	0.47	2.10
D	B	14.2	259	0.52	0.55	4.27
E	W	3.7	83	0.45	0.33	1.29
F	B	5.0	99	0.52	0.31	2.15

[a] After Antonis and Bersohn (14).
[b] W: White; B: Bantu.

diets, it was found that the latter group excreted less sterols and bile acids, when excretion was calculated on the basis of milligrams per gram dry weight of feces (18,19). When patients were placed on a liquid diet, fecal mass fell by 61% and fecal steroids by almost 90% (20). The first two of these studies were based on the subjects' own diets. In another experiment, patients were fed diets to which either bran or sugar cane bagasse had been added at a level of about 10 g/day (21). Both test substances resulted in bulkier feces. The daily excretion of neutral and acidic steroids was unaffected by bran but, when bagasse was fed, acidic steroid excretion rose by 50% and fecal neutral steroids fell by 10%. In contrast, another study showed that addition of 16 g/day of bran to the human diet resulted in increased stool weights but gave no significant changes in fecal bile acid levels (22). Addition of 33 g/day of bran to the diet has been reported to increase the chenodeoxycholic acid pool by 13.3 g at the expense of deoxycholic acid (23). Lignin, when added to the human diet, decreased bile acid half-life and increased the ratio of trihydroxy to dihydroxy bile acids (24).

The foregoing discussion illustrates how much must still be learned concerning the specific effects of individual types of fiber before conclusions are drawn.

When baboons were fed semipurified diets containing cellulose, they exhibited hyperlipidemia and aortic sudanophilia (25). The conversion of radioactive mevalonate to cholesterol in the bile of control animals was of the same order of magnitude as in the test animals. Conversion of mevalonate to bile acids in the controls was six times greater than in the animals fed the special diets. In addition, it was noted that the ratio of primary (cholic and chenodeoxycholic) to secondary (deoxycholic and lithocholic) bile acids in the control baboons was 1.67 and in the test animals it was 0.83 (average of four groups). These observations led to the hypothesis (26) that cholesteremia in the test animals was due to reduced elimination of bile acids, hence less conversion of biosynthesized cholesterol. In other words, in both dietary groups cholesterol synthesis continued at the same rate; in the controls there was greater excretion of bile acids; hence more conversion of cholesterol and less sterol was contributed to the circulation. The reverse was true in the test groups, i.e., less excretion of bile acid and, therefore, less conversion of cholesterol to bile acids with more diverted to the circulation. Kyd and Bouchier (27) reached a similar conclusion from a series of experiments involving cholelithiasis in rabbits. In an experiment in rabbits, similar results were obtained (28). Rabbits fed a semipurified diet containing cellulose exhibited cholesteremia and atherosclerosis whereas rabbits fed a commercial laboratory ration did not. The ratio of primary to secondary bile acids in the former group was twice that in the latter.

One possible mechanism of fiber action could involve the binding of bile salts. In this respect, dietary fiber would resemble the synthetic bile acid binding resins which are used as therapeutic agents for hypercholesteremia in man (29, 30). Eastwood and Hamilton (31) tested the binding *in vitro* of cholic and taurocholic acids to several fiber sources and found that a variety of affinities were affected by pH (Table IV). The wide range of binding capacities of those materials commonly lumped together under the designation "roughage" is evident. Birkner and Kern (32) also tested binding to foodstuffs. They used glycocholic and chenodeoxycholic acids. They found that potato had about 10 times the affinity for glycocholic acid that apple did. When tested against chenodeoxycholic acid, corn had almost twice the affinity of potato.

TABLE IV. Adsorption of Bile Acids to Fiber[a]

	Cholic acid		Taurocholic acid	
Materials[b]	pH 3.9	pH 8.0	pH 3.9	pH 8.0
Barley husk	0.97	1.05	1.05	0.75
Corn meal	1.10	1.28	0.74	1.25
Oak husk	0.77	0.51	0.38	0.40
Carrot	0.92	0.51	0.81	0.00
Turnip	0.91	0.31	0.67	0.30
Apple	1.00	0.59	1.14	0.53
Pear	1.00	0.59	0.90	0.58
Brussels sprout	1.09	1.03	1.33	1.18

[a] Eastwood and Hamilton (31).
[b] Bran mash = 1.00.

Kritchevsky and Story (33) compared the binding of taurocholate to materials commonly used as fiber in semipurified diets. Cellulose bound almost 30% less than did bran, which bound significantly less than did alfalfa. The binding capacity of alfalfa was striking. This substance bound a fifth of the amount of bile salt as did cholestyramine and a third as much as colestipol, two substances used clinically for lowering serum cholesterol levels. Balmer and Zilversmit (34) compared the binding *in vitro* of taurocholate to several types of grains and grain products. Their results are summarized in Table V. Cellulose bound no taurocholate under their conditions and lignin bound 2.1 times as much bile salt as did the stock diet.

Relatively few binding experiments had involved both tauro- and glycocholanoic acids. Since human bile contains three to four times as much glycocholate as taurocholate, experiments using both salts were carried out (35). It was found (Table VI) that the synthetic resins bound *in vitro* more taurocholate than glycocholate—not surprising since they had probably been screened against the former. The "natural" binding substances bound significantly more glycocholate. It was also noted that the three common bile acids were bound to alfalfa to different extents.

These findings prompted a larger study (36) in which the three common bile acids and their taurine and glycine conjugates were tested for binding *in vitro* to alfalfa, bran, cellulose, lignin, and cholestyramine. The results of this study are summarized in Table VII. It is immediately evident that each bile acid and each bile salt displays individual binding characteristics towards each binding substance. Among the "natural" materials, lignin exhibits the most consistently high levels of binding. In general, glycine conjugates are bound to a greater extent than are taurine conjugates.

TABLE V. Relative Binding of Sodium Taurocholate to Grains[a]

Binding substance[b]	Relative binding
Ground wheat	0.33
Soybean meal	0.69
Ground corn	0.80
Ground oats	0.65
Wheat middlings	0.95
Alfalfa meal	0.98
Lipid-extracted stock diet	0.90

[a] After Balmer and Zilversmit (34). Test substance (200 mg) incubated at 37°C for 30 min with 4 ml of micellar solution of bile salt.
[b] Ground stock diet = 1.00.

TABLE VI. Relative Binding of Sodium Tauro- or
Glycocholate to Various Substances[a]

Binding agent[b]	Taurocholate	Glycocholate
Alfalfa	1.00	1.13
Cholestyramine	3.47	2.89
Colestipol	2.97	2.62
Sugar beet pulp	0.11	0.15
Wheat straw	0.10	0.24

[a] After Kritchevsky and Story (35). 80 mg of binding agent and 100 μM
of bile salt in 5 ml saline (pH 7.0) incubated at 37°C for 1 h.
[b] Alfalfa/taurocholate = 1.00.

Among the free bile acids, chenodeoxycholate is generally more avidly
bound than deoxycholate. These data show that the structure of the bile
acid and of its conjugates determines the characteristics of its binding.
Since uncharacterized fibrous materials were used, more must be learned
about them in order to determine exactly the extent contributed by each
specific structural component towards the observed results.

In summary, various types of nonnutritive fiber have the capacity to
lower serum cholesterol levels, increase bile acid excretion, and decrease
bile acid turnover time in animals and in man. There are still several
important unknowns in this area. The first is the precise determination of

TABLE VII. Relative Binding of Bile Acids and Their Conjugates to Several
Substances[a]

Bile acids and conjugates[b]	Binding substance				
	Alfalfa	Bran	Cellulose	Lignin	Cholestyramine
Cholate (C)	1.00	0.51	0.02	2.20	3.05
Chenodeoxycholate (CDC)	1.25	0.91	0.10	1.17	4.35
Deoxycholate (DC)	0.52	0.27	0.01	0.87	4.64
Tauro C	1.00	0.20	0.14	11.70	3.20
Tauro CDC	2.19	1.42	0.00	13.83	3.68
Tauro DC	1.65	0.49	0.01	10.75	4.48
Glyco C	1.00	0.33	0.10	5.49	1.96
Glyco CDC	1.30	1.86	0.02	7.90	2.19
Glyco DC	2.42	0.68	0.41	7.62	4.57

[a] After Story and Kritchevsky (36). 50 mg of binding agent and 50 μM of bile acid or salt incubated in phosphate buffer (pH 7.0) for 2 h at 37°C.
[b] Binding of cholate or cholate salt to alfalfa = 1.00.

the type of dietary fiber being used. There are now methodologies (37–40) which permit determination of cellulose, hemicellulose, lignin, and other components of fiber. The second is the lack of understanding of how fiber exerts the effect that it does. Thus, bran, which has been highly recommended as a hypocholesteremic agent (41), has been shown to be ineffective in man (42–44) whereas guar gum (45) and pectin (45,46) lower cholesterol levels. There are also possible deleterious effects of excessive fiber intake (47) which must not be overlooked. The delineation of the mechanism(s) of action of various fiber types and of their constituents will provide a means of controlling bile acid metabolism and may contribute towards a better understanding of bile acid physiology.

REFERENCES

1. H. Trowell, *Proc. Nutr. Soc.* **32**, 151 (1973).
2. D. P. Burkitt, A. R. P. Walker, and N. S. Painter, *J. Am. Med. Assoc.* **229**, 1068 (1974).
3. G. A. Spiller and R. J. Amen, *Am. J. Clin. Nutr.* **28**, 675 (1975).
4. J. A. Story and D. Kritchevsky, *in* "Fiber in Human Nutrition" (G. A. Spiller and R. J. Amen, eds.), p. 171, Plenum, New York (1976).
5. O. W. Portman and G. V. Mann, *J. Biol. Chem.* **213**, 733 (1955).
6. O. W. Portman, G. V. Mann, and A. P. Wysocki, *Arch. Biochem. Biophys.* **59**, 224 (1955).
7. O. W. Portman and P. Murphy, *Arch. Biochem. Biophys.* **76**, 367 (1958).
8. O. W. Portman, *Am. J. Clin. Nutr.* **8**, 462 (1960).
9. G. A. Leveille and H. E. Sauberlich, *J. Nutr.* **88**, 209 (1966).
10. W. T. Beher and K. K. Casazza, *Proc. Soc. Exp. Biol. Med.* **136**, 253 (1971).
11. D. Kritchevsky, R. P. Casey, and S. A. Tepper, *Nutr. Rep. Int.* **7**, 61 (1973).
12. D. Kritchevsky, S. A. Tepper, and J. A. Story, *Nutr. Rep. Int.* **9**, 301 (1974).
13. H. Fisher and P. Griminger, *Proc. Soc. Exp. Biol. Med.* **126**, 108 (1967).
14. A. Antonis and I. Bersohn, *Am. J. Clin. Nutr.* **11**, 142 (1962).
15. A. R. P. Walker, B. F. Walker, and B. D. Richardson, *Br. Med. J.* **3**, 48 (1970).
16. G. O. R. Holmgren and J. M. Mynors, *S. Afr. Med. J.* **46**, 918 (1972).
17. D. T. Forman, J. E. Garvin, J. E. Forestner, and C. B. Taylor, *Proc. Soc. Exp. Biol. Med.* **127**, 1060 (1968).
18. V. C. Aries, J. S. Crowther, B. S. Drasar, M. J. Hill, and F. R. Ellis, *J. Pathol.* **103**, 54 (1971).
19. M. J. Hill and V. C. Aries, *J. Pathol.* **104**, 129 (1971).
20. J. S. Crowther, B. S. Drasar, P. Goddard, M. J. Hill, and K. Johnson, *Gut* **14**, 790 (1973).
21. R. L. Walters, I. M. Baird, P. S. Davies, M. J. Hill, B. S. Drasar, D. A. T. Southgate, J. Green, and B. Morgan, *Br. Med. J.* **2**, 536 (1975).
22. M. A. Eastwood, J. R. Kirkpatrick, W. D. Mitchell, A. Bone, and T. Hamilton, *Br. Med. J.* **4**, 392 (1973).
23. E. W. Pomare and K. W. Heaton, *Br. Med. J.* **4**, 262 (1973).
24. K. W. Heaton, S. T. Heaton, and R. E. Barry, *Scand. J. Gastroenterol.* **6**, 281 (1971).

25. D. Kritchevsky, L. M. Davidson, I. L. Shapiro, H. K. Kim, M. Kitagawa, S. Malhotra, P. P. Nair, T. B. Clarkson, I. Bersohn, and P. A. D. Winter, *Am. J. Clin. Nutr.* **27**, 29 (1974).
26. D. Kritchevsky, S. A. Tepper, and J. A. Story, *J. Food Sci.* **40**, 8 (1975).
27. P. A. Kyd and I. A. D. Bouchier, *Proc. Soc. Exp. Biol. Med.* **141**, 846 (1972).
28. D. Kritchevsky, S. A. Tepper, H. K. Kim, D. E. Moses, and J. A. Story, *Exp. Mol. Pathol.* **22**, 11 (1975).
29. S. A. Hashim and T. B. Van Itallie, *J. Am. Med. Assoc.* **192**, 289 (1965).
30. J. R. Ryan and A. Jain, *J. Clin. Pharmacol.* **12**, 268 (1972).
31. M. A. Eastwood and D. Hamilton, *Biochim. Biophys. Acta* **152**, 165 (1968).
32. H. J. Birkner and F. Kern, Jr., *Gastroenterology* **67**, 237 (1974).
33. D. Kritchevsky and J. A. Story, *J. Nutr.* **104**, 458 (1974).
34. J. Balmer and D. B. Zilversmit, *J. Nutr.* **104**, 1319 (1974).
35. D. Kritchevsky and J. A. Story, *Am. J. Clin. Nutr.* **28**, 305 (1975).
36. J. A. Story and D. Kritchevsky, (Unpublished Results).
37. P. J. Van Soest and R. W. McQueen, *Proc. Nutr. Soc.* **32**, 123 (1973).
38. I. M. Morrison, *J. Sci. Food Agric.* **23**, 455 (1972).
39. C. S. Edwards, *J. Sci. Food Agric.* **24**, 381 (1973).
40. D. A. T. Southgate, *J. Sci. Food Agric.* **20**, 331 (1969).
41. D. Reuben, "The Save Your Life Diet," Random House, N.Y. (1975).
42. M. A. Eastwood, *Lancet* **2**, 1222 (1969).
43. K. W. Heaton and E. W. Pomare, *Lancet* **1**, 49 (1974).
44. A. M. Connell, C. L. Smith, and M. Somsel, *Lancet* **1**, 496 (1975).
45. D. J. A. Jenkins, A. R. Leeds, C. Newton, and J. H. Cummings, *Lancet* **1**, 116 (1975).
46. A. Keys, F. Grande, and J. T. Anderson, *Proc. Soc. Exp. Biol. Med.* **106**, 555 (1961).
47. M. M. L. Sutcliffe, *Br. J. Surg.* **55**, 903 (1968).

AUTHOR INDEX

SUBJECT INDEX